Applied Cyber-Physical Systems

Sang C. Suh · U. John Tanik
John N. Carbone · Abdullah Eroglu
Editors

Applied Cyber-Physical Systems

Foreword by B. Thurasingham

 Springer

Editors

Sang C. Suh
Department of Computer Science
Texas A&M University—Commerce
Commerce, TX
USA

John N. Carbone
Raytheon Intelligence and Information
  Systems
McKinney, TX
USA

U. John Tanik
Department of Computer Science
Indiana University—Purdue University
  Fort Wayne
Fort Wayne, IN
USA

Abdullah Eroglu
Department of Electrical Engineering
Indiana University—Purdue University
  Fort Wayne
Fort Wayne, IN
USA

ISBN 978-1-4899-8854-6   ISBN 978-1-4614-7336-7 (eBook)
DOI 10.1007/978-1-4614-7336-7
Springer New York Heidelberg Dordrecht London

Printed on acid-free paper

Springer is part of Springer Science+Business Media (www.springer.com)

Foreword

Cyber-Physical Systems are those systems that integrate computing systems with the physical components. These include a wide variety of systems such as aerospace systems, command and control systems, manufacturing systems, process control systems, robotic systems, telecommunication systems, power grid systems, and biomedical systems. Cyber-Physical systems have infiltrated almost every field of engineering including chemical, electrical, petroleum, power, mechanical, civil, biomedical, and aerospace engineering. Until recently the cyber component and the physical component were not seamlessly integrated. However, due to pervasive computing in almost every aspect of our lives, there is now an urgent need to integrate them. Furthermore, the advent of the cloud and service computing models has contributed significantly to the growth of cyber-physical systems. Large-scale industrial automation systems now use the power of the cloud to carry out massively parallel computations.

Several computing technologies have to work together to build efficient cyber-physical systems. At the heart of cyber-physical systems are the embedded processors. Numerous microprocessors are now integrated into the physical systems such as our homes and our cars. These microprocessors need special types of networking and data management protocols. For example, the data management systems have to operate in main memory with caches and therefore need novel query and storage capabilities. Second, wireless computing devices and smartphones have to be tightly integrated with the physical components requiring additional challenges for data management. Cloud computing and service-oriented models require novel technologies for virtualization and resource utilization.

Last but not least, security and privacy are major considerations for cyber-physical systems. We hear of reactively adaptive malware attacks on our critical infrastructures such as our power grids and smart meters, sometimes several times a day. These powerful malware can change their patterns including their behavior rapidly. Therefore, current antivirus products cannot detect such malware. We need solutions for not only detecting such malware but also for recovering from such malware attacks so that the mission can be carried out on time. Powerful machine learning techniques are being explored to learn the behavior of the adversary so that effective solutions can be developed to thwart the adversary.

Furthermore, appropriate controls for preventing such attacks are needed. Finally, we need solutions to maintain the privacy of individuals. For example, through the use of smartphones connected to cyber-physical systems one can find out private information such as the location of the individual. We need privacy preserving techniques for cyber-physical systems. The solutions include novel cryptographic protocols as well as secure privacy enhanced wireless networking and data management techniques.

While research programs are under way at the National Science Foundation and the Department of Defense, among others, to promote and advance cyber-physical systems research, we need tighter integration with the industries that are developing the physical components such as the defense, automotive, and aerospace industries. Therefore we need strong partnerships among the government, industry, and academia to develop highly innovative and effective solutions for cyber-physical systems.

While there have been various articles and books on cyber-physical systems, this is the first book of its kind to provide a comprehensive overview of the developments in cyber-physical systems. The papers include architectures, models, and infrastructures for cyber-physical systems, security and privacy for cyber-physical systems, applications of cyber-physical systems and design methods, tools and components for cyber-physical systems. The in-depth articles will not only enhance the research in the field, but will also provide approaches and solutions to some of the urgent problems faced by practical cyber-physical systems. The book will also help curriculum developers with guidance to develop courses in the field. It will provide the foundations for establishing both graduate and undergraduate research as well as education programs in the field. Finally, it will contribute to the increasingly important area of cyber operations with respect to cyber-physical systems.

The urgent need to securely integrate the cyber and physical components of a system make this book a must read for anyone working in the computing and engineering fields.

Dr. Bhavani Thuraisingham
Louis A. Beecherl Jr., Distinguished Professor
Department of Computer Science
Executive Director of the Cyber Security Research
and Education Center
Erik Jonsson School of Engineering and Computer Science
The University of Texas at Dallas
Dallas, TX, USA
E-mail: bhavani.thuraisingham@utdallas.edu
http://www.utdallas.edu/~bxt043000/

Preface

This is the second book in the SDPS book series. Our first book in the series was *Biomedical Engineering: Health Care Systems, Technology and Techniques*. This second book in the series titled *Applied Cyber-Physical Systems* is developed by distinguished editors Dr. John Carbone, Dr. John Urcun Tanik, and Dr. Abdullah Eroglu. Each of these editors has specific expertise in an aspect of Cyber-physical systems. In addition, they were instrumental to receive the endorsements of two senior experts in the area. Dr. Bhavani Thuraisingham, an international expert on cyber security and Dr. Bernd Krämer, an international expert in the early applications of cyber-physical systems.

Generally speaking, the kind of topics SDPS book series publishes is actually defined for us by Nobel Laureate, late Herbert Simon during the SDPS-2000 Keynote Speech, *We have learned very well that many of the systems that we are trying to deal with in our contemporary science and engineering are very complex indeed. They are so complex that it is not obvious that the powerful tricks and procedures that served us for four centuries or more in the development of modern science and engineering will enable us to understand and deal with them. **We are learning that we need a science of complex systems, and we are beginning to construct it**...*

Society for Design and Process Science, since 1995, is striving to develop scientific foundations for design and process to address issues related to complex systems, while maintaining disciplinary integrity of its membership. Therefore, we promote books addressing complex problems, which are not admissible to single discipline solutions. Since our membership comes from diverse disciplines, from mathematics to medicine, we do not offer any disciplinary boundary to our book series. For example, our 2010 transformative achievement award winner Dr. Steven Weinberg is a Nobel Laureate in Physics, while the first winner of the same medal from SDPS (2000) is late Herbert Simon who was a Nobel Laureate in Economics. On the other hand, SDPS 2011 transformative achievement awards went to the developer of Fuzzy Sets Dr. Lotfi Zadeh, a computer scientist from

Berkeley and the developer of Axiomatic Design Theory Dr. Nam Suh, who is a distinguished engineer and president of KAIST research center in Korea, while SDPS 2012 transformative awards goes to Dr. Edward O. Wilson of Harvard University and Dr. Bhavani Thuraisingham of University of Texas at Dallas.

Being in line with SDPS goals and vision, this second book in the series aims to provide the cutting edge methods and technologies in the area of Cyber-physical systems. The notion of cyber-physical systems constitutes the backbone for a future cyber-physical society in which not only the cyber and physical spaces but also humans (physiological and psychological spaces), knowledge, society, and culture is incorporated. Cyber-physical systems integrate computing and communication capabilities by monitoring and controlling the physical systems via embedded hardware and computers.

The purpose of writing the book is to provide the readers with an insight into the transformative and trans-disciplinary research conducted by the members of the Society for Design Process Science (SDPS) in the broad areas of CPS. A main challenge in CPS when deployed for any application is the lack of a unifying framework that integrates all the recent and expected advances made in embedded systems development in order to maintain intellectual control over System of Systems (SoS) deployments. This book will address these issues by providing various diverse applications as examples.

We congratulate the distinguished editors, for doing an excellent job by bringing together such a diverse expertise in the broad area of Cyber-physical systems.

Dr. Sang Suh and Dr. Murat M. Tanik
SDPS Series Editors

This volume focuses on the upcoming near-term, mid-term, and long-term challenges we all face in the onslaught of rapidly developing Cyber-Physical Systems within our current century and beyond. The papers are focused upon providing insights from global subject matter experts within the industry and academia. The papers are organized into groups, which take the reader through important background information, new concepts, methods, and solutions for improving education, as well as, concepts for changing the way we think, perceive problems, and achieve the resulting solutions. In inviting subject matter experts to provide insights we have specifically taken into consideration diverse topics focusing on papers which discuss practical research and solutions relating to designs, implementations, and reliability of Cyber-Physical Systems.

Without the dedicated global participation of diverse authors this book would not have been possible. Society for Design and Process Science (SDPS) conferences have played a pivotal role in providing the CPS subject diversity of thought

and the acquisition of expanded versions of compelling conference papers for this
volume. We hope this book provides practicality and motivation for continued
research in this area and we look forward to your thoughts and debate.

Sang C. Suh
U. John Tanik
John N. Carbone
Abdullah Eroglu

Contents

Chapter 1
Evolution of Cyber-Physical Systems: A Brief Review

Bernd J. Krämer

The term "cyber-physical system" (CPS) sounds like a brand-new buzzword as it occurs increasingly as a theme of many conferences, in journal articles and books – like this one. Etymologically the prefix cyber derives from the ancient Greek word κυβέρνησις (kybernesis) and originally means control skills. It evolved into the Latin word *guvernare* and finally the English word to *govern*. For our context this means that we speak about systems in which physical objects and computational resources are tightly integrated and exhibit a degree of continuous coordination between each other.

Sztipanovits characterizes cyber-physical systems research "as a new discipline at the intersection of physical, biological, engineering and information sciences" [1]. In this broad sense, Konrad Zuse was a pioneer in cyber-physical systems. Why? Soon after the invention of the Z3 in 1941, the first fully functional program controlled computation machine, he developed a special device for the survey of aircraft wings. Zuse later called this ensemble the first real-time computer. This automatic computer read values from some forty sensors, working as analogue-to-digital converters, and processed these values as variables within a program. We conclude from this that real-time capabilities, reactivity, control engineering, software, and physical resources are inherent aspects of cyber-physical systems.

Somewhat later, in 1948, Norbert Wiener coined a new term in his book *Cybernetics: or Control and Communication in the Animal and the Machine* in which he elaborated on feedback concepts between men and machines, including feedback mechanisms in technical, biological and social systems. A second addition of this book, which appeared in 1961, showed how long-sighted Wiener was because he added two new chapters *On Learning and Self-Reproducing Machines* and *Brain Waves and Self-Organizing Systems*. These are still popular topics in research, for instance: self-reproduction in the context of outer space exploration or Nano technology; self-organization in bio-inspired multi-agent

B. J. Krämer (✉)
Scientific Academy for Service Technology, Hagen, Germany
e-mail: Kraemer@servtech.info

S. C. Suh et al. (eds.), *Applied Cyber-Physical Systems*,
DOI: 10.1007/978-1-4614-7336-7_1,
© Springer Science+Business Media New York 2014

1

models or distributed flight areas as Raffaello d'Andrea demonstrated them in his keynote during the SDPS 2012 conference in Berlin. Long before the Society for Design and Process Science (SDPS) with its mission in transdisciplinary research and education was founded, cybernetics conferences in the fifties insisted on a finely balanced ratio of representatives from different disciplines, including mathematics, sociology, physiology and more (cf., e.g., [2]).

Claus Pias nicely characterizes the current revival of cyber-* when he says [3]: "Today the cyber-hype of these days appears like a fashion garment of yesterday, and the threshold of shame to wear it, is low anytime". Shame is not appropriate anyway because the prerequisites we find today are far more advanced than 50 years ago. We have built up a rich body of real-time and embedded systems engineering know-how; we have the world-wide web that evolved from a web of information over a web of software services to a web of people and organizations; we find advanced sensor and actuator technology and we have created a tremendous amount of algorithmic and software engineering knowledge. The chances to experience the benefits of modern cyber-physical systems research steeply increased: We see the realization of remote and robot-enabled surgery on the horizon. The Internet of things is subject to substantial research efforts and funding. Car manufacturers start embedding vehicle collision prevention systems into premium cars; the realization of low-energy buildings or renewable energy sources seamlessly integrated into smart power grids has top priority on the political agenda, even more so in public awareness. The European Union is investing substantial funds in the research and development of smart factories. For instance, the EU-project IMAGINE, which organized a workshop on End-to-end management of dynamic manufacturing networks during SDPS 2012, presented the project's approach towards the effective (re-)configuration and management of complex dynamic manufacturing networks.

This brings us to a sharper definition of cyber-physical systems by adding further characteristics: Many of the application areas mentioned are inherently *distributed* and equipped with *wire-bound or wireless communication* facilities. Their components are largely autonomous and need to be coordinated and controlled. A CPS used in such critical domains as dynamic and prospective traffic safety, factory and process control, or healthcare needs to be *highly dependable* requiring the availability of reliability performance, availability, safety and security.

Of course, the European Commission is not the only institution placing its hopes in the promotion of the next generation of cyber-physical systems. The US National Science Foundation (NSF) supported this type of research already since 2006. It funded, for example, projects on sensor-based autonomous systems, distributed robotics, autonomous vehicles (both land and air), and ambient assisted living. Encouraged by a recommendation in the December 2010 report of the President's Council of Advisors on Science and Technology on Designing a Digital Future, NSF will soon open a new window for proposals on new cyber-physical systems projects. Between mid 2010 and January 2012 the German Federal Ministry of Education and Research and German industry set up a cooperative study project to develop an "Integrated Research Agenda Cyber-Physical Systems". The consortium involved major German industries and research institutes and investigated technological, economical, political, and social challenges and impacts of technology trends towards cyber-physical systems [4].

Besides further research in different application domains, a fresh look at CPS also requires a new transdisciplinary engineering approach. As we speak about hybrid systems including electronics, mechanics, software and other technical components, new approaches towards integrated systems modeling approach, a coherent design theory and related design, analysis and simulation tools become indispensable. However, cyber-physical systems are not just a self-contained and isolated ensemble of technical components but are often embedded in a social context to form a socio-technical system. In such systems people are embedded in complex organizational structures and interact with complex infrastructures to perform their work processes. A holistic approach towards human factors, including usability of interfaces and functionality, intuitive machine operating, and seamless coordination of human and machine behavior are of outmost importance to avoid erroneous system behavior.

Correspondingly, new curricula addressing relevant fields of knowledge need to be established at different levels of education. The Society for Design and Process Science has pioneered transdisciplinary engineering education, but CPS requires further efforts, in particular, with respect to innovative design disciplines for man–machine interfaces and interaction and the seamless embedding of CPS in the actual application domain. We also need to understand and teach how to incorporate adaptivity and context awareness in CPS.

The authors of this book touch many of the issues mentioned above in great detail and with high scientific and practical competencies. They step onto new grounds and shed a light on many aspects of our ignorance in this challenging field. I'm overly grateful to the editors of this book that they undertook the courageous endeavor of characterizing this technological field, dividing it into meaningful subthemes, and finding outstanding authors for each chapter.

References

1. Sztipanovits, Janos (2007). 14th Annual IEEE International Conference and Workshops on the Engineering of Computer-Based Systems (ECBS '07), pp. 3–6, IEEE Computer Society
2. Pias, Claus (Ed., 2003): Die Macy-Konferenzen (the Macy Conferences) 1946–1953. Berlin: Diaphenes (in German)
3. Pias, Claus (2003): Zeit der Kybernetik – eine Einstimmung (Time of Cybernetics—To Get Attuned). In [2], pp. 9–41 (in German)
4. Geisberger, Eva, Broy, Manfred (Eds., 2012): agendaCPS—Integrierte Forschungsagenda Cyber-Physical Systems (in German), Springer, series acatech Studie

Chapter 2
The Need for a Transdisciplinary Approach to Security of Cyber Physical Infrastructure

Jim Brodie Brazell

Introduction

Cyber-physical systems are an emerging domain of warfare, terrorism, and crime, while simultaneously being the platform and engine for economic, civil, and democratic development in the twenty first century. The protection and sustainable growth of cyberspace is necessary for advancement of civilization and the pursuit of the noble goals of humanity. However, it has become common in the last decade to learn of cyber exploitations that threaten the safety and security of daily life as we know it.

Cyber exploitation can result in information theft, violation of privacy, disruption of economic services, and even physical effects projected from cyberspace to the natural world. Cyber attacks are not confined to the realm of cyberspace. A cyber attack can also inhibit, surveil, or damage physical property such as machines, motors, and physical processes controlled by computers negatively impacting sustainable economic growth and progress.

On March 11, 2011, the New York Times reported that researchers were able to hack a car remotely and take control of critical systems via embedded communications systems in the car [1]. Because many of today's cars contain cellular connections and Bluetooth wireless technology, it is possible for a hacker, working from a remote location, to hijack various features, such as the car door locks and brakes, as well as to track the vehicle's location, eavesdrop on its cabin, and monitor vehicle data. This hack demonstrates how cyberspace can be used to affect physical processes beyond cyberspace.

Late in the twentieth century, the physical infrastructure of transportation, energy, water, and industrial systems were integrated with embedded digital computing and network communications. Thus, processes controlling cars, water filtration, oil refineries, electrical grids, and even consumer electronics have

J. B. Brazell (✉)
Ventureramp, Inc, 9515 S. Saddle Trail, San Antonio, TX 78255, USA
e-mail: Jimbrazell@ventureramp.com

S. C. Suh et al. (eds.), *Applied Cyber-Physical Systems*,
DOI: 10.1007/978-1-4614-7336-7_2,
© Springer Science+Business Media New York 2014

changed from mechanical and analog systems to robotic-like devices that integrate digital communications and control systems.

The category of computing in embedded systems and large scale distributed computing applications that link physical processes and digital control is called cyber-physical systems. Cyber-physical systems constitute both legacy systems such as supervisory control and data acquisition (SCADA), and the emergence of new, massively distributed intelligent systems constituting a 4th generation of computing (cyber-physical systems and infrastructure). Modern and legacy "cyber-physical" systems are brittle and vulnerable to cyber exploits in the same way that computers and networks are open to attack.

Most legacy cyber-physical systems were *not* designed for security. Recent examples of security breaches include the Stuxnet and Shamoon viruses. Stuxnet disrupted Siemens' programmable logic controllers integral to Iranian uranium enrichment processes. Shamoon hit Saudi Arabian state oil company Aramco, and Qatari natural gas producer RasGas, rendering 30,000 computers useless and constituting what Leon Panetta labeled the most destructive cyber attack the private sector has seen to date [2].

Although Shamoon did not attack physical control systems like Stuxnet, the malware's success at stuttering administrative computer systems in the oil fields of Arabia points to the imminent threat of disrupting cyber-physical systems as a next step. For example, cyber spies penetrated the U.S. electric grid and deposited software tools that could be used to damage or disrupt the grid. National Security forensics indicate that the spies came from China, Russia, and other countries [3]. (Fig. 2.1).

Cyber-physical systems are the product of a transdisciplinary engineering design process–mechatronics–that integrates electronic, software, computer, and motor control. Cyber-physical systems and transdisciplinary design are therefore important to the security of these increasingly integrated and pervasive systems. Transdisciplinary research exists at the intersection of disciplines necessary to promote transformation and advancement of technical, organizational, and social systems through process and design science.

Fig. 2.1 Emerging technology: 4th generation computing

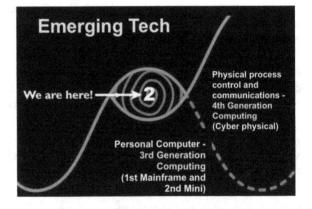

The Concept of Cyber-Physical Systems

The infrastructure enabling the basic services of life in the twenty first century are now governed by and dependent on cyber physical control systems. Business, military, government, and civil society now depend on the integrity and resilience of inter-networked computers and remote control physical processes (cyber physical infrastructure).

The United States' Executive Order 13010 defines "infrastructure" as the framework of interdependent networks and systems comprising identifiable industries, institutions (including people and procedures), and distribution capabilities that provide a reliable flow of products and services essential to the defense and economic security of the United States, the smooth functioning of government at all levels, and society as a whole [4].

The private sector owns and operates an estimated 85% of infrastructure and resources that are critical to our nation's physical and economic security [5]. Cyber-physical systems are at the heart of the design, manufacture, installation, operation, and sustainability of private corporate and national infrastructure across the world, yet these systems are highly vulnerable to cyber exploitation. This cyber-physical infrastructure includes the physical centralized plants (such as manufacturing and processes plants), the physical distributed infrastructure (such as electrical grids and pipelines), and even the disaggregated vehicles that consume and transport the products of these systems (such as trucks, planes, and ships).

> *Definition*–Cyber-physical systems utilize information technology (computers, software, and networks) to direct the communication and control of physical processes and systems (or vice versa). Cyber-physical systems include legacy analog and digital systems such as supervisory control and data acquisition (SCADA), machine-to-machine (M2M) computing, industrial control, and generally, all embedded systems utilizing automatic control techniques.

Examples of cyber-physical systems span the cochlear ear implant, heart pacemaker defibrillator, anti-lock brakes on vehicles, computerized control of gas pipeline valves, computerized control of weapons systems, computerized control of manufacturing and refinery processes, and computerized control of pumps on dams and motors on drawbridges. Cyber-physical systems are quickly expanding in application and adoption across public and private sectors.

Cyber-physical systems include home appliances and even digital mobile phones that remotely interface with home automation systems, or automobiles. New regulatory policy, technology standards, business models, and risks are emerging as cyber-physical systems advance with increasing power and decreasing cost. By one estimate, Tesar's Law, tightly coupled cyber and physical systems (intelligent actuators) have exhibited an eight-fold ($8\times$) price/performance increase during the past 20 years to parallel Moore's Law for computer processors [6]. The performance characteristics of cyber-physical systems represent efficiencies to be leveraged by industry to gain greater economic performance through automation. During the "great recession", companies have exercised investment to increase automation, thereby ushering in a new era of automation characterized by greater diffusion of cyber physical control systems that are transparent to many of the services they provide.

In the past, industrial control, physical security, cyber security, environmental control, consumer products, automotive, heavy equipment, and defense systems have existed as separate and distinct industries and markets with relatively distinct infrastructures and technological systems.

Today, the underlying technologies of cyber-physical systems are converging, in the context of security and technology architecture, due to the de facto acceptance of standardized, modular hardware and software, as well as the acceptance of Internet protocol as an underlying method of communications in many of these systems. Legacy cyber-physical systems and many emerging systems, however, are not designed for security, nor were many cyber physical processes designed to be connected to open network systems and processes such as the Internet. The prevalence and nature of cyber physical control systems at the heart of infrastructure necessitate a shift in thinking about these systems from a physical security perspective to a systems design- and process-centered view.

The design of cyber-physical systems rests on the foundation of the theory of cybernetics developed by Norbert Weiner at MIT in the 1950s; the practice of mechatronic design fostered in 1969 by Tetsuro Mori, a senior engineer at the Japanese motion control company Yaskawa; and the emergence of design and process science in 1995 fostered by Dr. C. V. Ramamoorthy, Dr. Raymond Yeh, Dr. Murat Tanik, and Dr. George Kozmetsky (1917–2003) and others from the Society for Design and Process Science (Fig. 2.2).

Fig. 2.2 Rensselaer Polytechnic Institute mechatronics model [7]

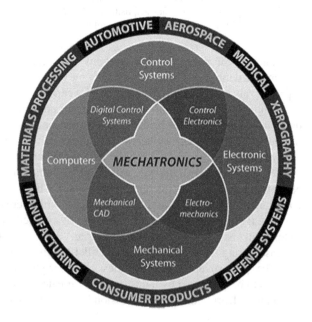

As an area of research, cyber-physical systems design and sustainability span many industries, technology domains, and educational arenas. For example, engineering design and manufacturing installation, operation, and maintenance depend on the consilience of mechanical, electronic, computer, and software domains traditionally associated with "mechatronics." Cyber-physical systems expand the reach of automation and have a workforce effect of decreasing human labor input in real, full-time equivalency, while requiring an "up skilling" of equipment operators, technicians, engineers, and scientists characterized as multi-disciplinary, multi-skill, and transdisciplinarity [8–10]. This requirement for transdisciplinarity is at the heart of what constitutes the next generation science, technology, engineering and mathematics (STEM) workforce.

Transdisciplinarity is rooted in design and process science. As such, rather than being a discipline, transdisciplinarity represents the aggregation of knowledge and human capability within the context of solving unstructured problems for which there are no answers. It is in effect, the coalescence of the necessary resources and human skill to design and create what is next.

For example, there is a deficit in workforce and academic education programs relative to cyber-physical systems and security. While mechanical engineering programs are often delivered at the university level without much focus on digital communications and control theory, most mechanical systems such as cars, trains, planes, satellites, rockets, washing machines, clothes dryers, automatic dish-washing machines, water sprinkler systems, and even many toys now integrate digital communication and control of mechanical systems and processes. These systems are often designed without a preponderance of security, or the reality that essentially all machines are now effectively extensions of a global communications network.

Transdisciplinary research can help to develop secure solid-state chips and components to meet commercial and military application capability standards. New standards can be defined for cyber-safe components similar to ID-approaches used by today's computer central processing units (CPU) [11]. We have reached a time when all devices and equipment need to be designed from the beginning for interaction with networks and for security from the component-level, to the systems-level.

Sustainability of cyber physical infrastructure requires a wider systems design and process science approach to applications such as human–machine cooperation. From an academic perspective, cyber-physical systems security touches disciplines concerned with the social sciences and the field of convergent science, engineering, information technology, and life science. From a systems perspective, cyber-physical systems tend to unify information, communications, and crypto-logical systems theory in the context of hardware and software design integrating human, biological, chemical, nuclear, ultrasonic, sonic, gravity, magnetic, optical, radio frequency, electro-magnetic, and x-ray fields of study and application [12] (Fig. 2.3).

The increased reliance on cyber physical control systems for critical infra-structure means that a security process approach is required in addition to

Fig. 2.3 National Science
Foundation: changing the
societal "fabric" towards a
new structure [13]

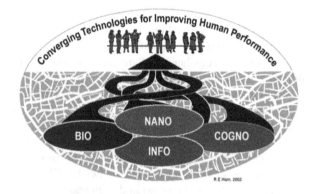

prevalent asset protection schemes. The function of developing new procedural
approaches to the design, manufacture, implementation, and sustainability of
cyber-physical systems requires multi-sector coordination. The unmet need in the
realm of cyber-physical systems security is the creation of processes that can
effectively enable multi-sector collaboration and governance in the interest of the
whole, without sacrificing the autonomy of individual actors.

Cyber Physical Workforce Need

Cyberspace is the platform upon which U.S. and global wealth creation has been
built since the Apollo program—spanning virtually all forms of business, engineering, life science, physical science, and social science occupations as well as
copyright industries (arts industries).

According to the President's Council of Advisors on Science and Technology
(PCAST) report Designing Our Digital Future, increasing the number of graduates
in [Network and Information Technology (NIT)] fields at all degree levels must be
a national priority. Furthermore, the PCAST report states, "All indicators—all
historical data, and all projections—argue that NIT is the dominant factor in
America's science and technology employment, and that the gap between the
demand for NIT talent and the supply of that talent is and will remain large" [14].

According to the Bureau of Labor Statistics, NIT is projected to account for
between 52 and 58 % of all STEM occupations between 2008 and 2018 with
projected NIT employment resulting in 762,700 jobs, growing more than twice as
fast as the average for all occupations in the economy [15].

Emerging requirements for critical infrastructure protection indicate a need for
technical training and advanced component certification for cyber security professionals as well as for technicians who work at the intersection of cyber and
physical systems (e.g. industrial maintenance, process control technicians, and
instrumentation technicians). In addition there is a specific unmet need for a body

of professionals who have trained to work together across public and private sectors to address cyber physical incidents cooperatively.

The technicians who work at the intersection of cyber and physical systems and processes are referred to as "multi-craft technicians," "multi-skill technicians," "integrated systems technicians," "mechatronics technicians," "engineering technicians," "instrumentation and control systems technicians," "process control technicians," and "Science and Technology R&D technicians," among other names. In workforce education, there are a few systems technician programs; however, little systematic focus has been placed on cyber-physical systems security in the context of distributed critical infrastructure protection, classical cyber security training, and/or education for technicians who operate at the interface of cyber systems and physical systems, but who are not classically considered information technology or security workforce personnel.

There are critical gaps in the daily management of cyber security incidents within institutions and among institutions (public and private). In the emergency response community, systems processes and operational procedures do not typically include or integrate cyber into operational readiness on-site when in effect the whole response is coordinated by cyber communications and computing infrastructures. This gap in the Emergency Center Operations protocol and capability puts the nation at risk to both physical and virtual attacks by cyber adversaries, and sets up a self-imposed information asymmetry among friendly cyber security actors.

The Center for Infrastructure Assurance and Security (CIAS) at the University of Texas San Antonio (UTSA) has received funding through the Department of Homeland Security (DHS) to conduct a series of training classes, workshops, and exercises in communities across the country. The program that is being followed is based on the Community Cyber Security Maturity Model (CCSMM) developed by the CIAS. The CCSMM came about as a result of what the CIAS observed as it conducted cyber security exercises between 2002 and 2005.

CIAS personnel work with communities to organize and conduct cyber security exercises. After several years, the CIAS observed that when they returned to one of the communities in which they had conducted an exercise that very little, or nothing, had been done to improve the community's security posture even though the leaders were aware of the potential impact of a cyber incident. As a result, the CIAS took a hard look at what was missing and developed the CCSMM [16].

The CCSMM provides three main functions. First, it provides a "yardstick" for a community to measure where it is in the maturity of its cyber security program. Second, it provides a "roadmap" to show a community what is needed and what steps to take in order to improve its posture from where it currently is. Finally, it serves as a common point of reference for individuals from different communities to communicate about their respective programs. As the model was developed, training courses were identified that could help communities and, where no courses existed, the CIAS, through a separate DHS grant, developed courses for states and communities. The result was a phased plan and approach for communities to follow to improve their security posture [ibid].

In addition to the training, workshops, and exercises the CIAS identified as part of the model, other necessary elements of a viable and sustainable security program were also identified. One of the identified needs was the necessity for every community to have a group of trained individuals who could work together in order to address a cyber incident. In addition, community emergency management leaders needed to understand the basics of cyber incident response and how to manage a cyber incident affecting their community. The problem is that, while there are a number of individuals who know something about cyber security, there are very few (if any) individuals in a community who understand how to handle a cyber incident from a multi-institutional coordination perspective. A primary mission of the CCSMM and CIAS is to address this human and institutional communication and collaboration need for cooperative incident management [ibid].

Malicious cyber attacks are an everyday occurrence in twenty first century cities, enterprises, and even the public and private lives of global citizens. Good practice and preparedness can go a long way, but we must inevitably be prepared for the lights to go out and for insuring the availability of multi-sector coordination and specialized human talent required to handle emergency cyber incidents, including the full spectrum of operational processes enabling not only response but also proactive defensive measures: coordination, prevention, protection, mitigation, and recovery. In the end, we are all first responders; therefore, cyber mitigation includes professional and technical workforce preparedness as well as the necessary education for all students and citizens to be prepared and responsive to the security and safety requirements of cyberspace.

Conclusion

Preliminary analysis indicate an overall lack of understanding of what to do in response to cyber physical attacks, as well as how to create a state of operational readiness and preparedness for major cyber and/or dual cyber and physical events. Given the escalation of current cyber penetration and related havoc, a general requirement exists for proactive measures in addition to reactive measures in cyber security and incident coordination management in the domain of cyber-physical systems.

A shortcoming of existing critical infrastructure protection is to organize primarily using security policies and procedures based on the perimeter defense model to protect infrastructure as a physical asset. The increasing ubiquity of cyber systems that govern and mediate the physical processes of infrastructure necessitate a coordinated approach to cyber-physical systems security, workforce education, and incident response affecting research and development, training and certification, and emergency response protocol.

The systemic nature of cyberspace and the nature of cyber exploitation indicate a need for new organizational and disciplinary approaches to how institutional actors organize and affect security operations. Advancement of cooperative incident management in cyber-physical systems security requires a transdisciplinary approaches to both academic and real world operational needs.

Transdisciplinary approaches to cyber-physical systems present opportunities for enhanced security of critical infrastructure and related economic, political, and civic sustainability. Cyber-physical domains of operation include critical infrastructure: emergency services, tele-communications, energy, law enforcement, fire departments, utilities, public works, medical, industrial facilities, banking and finance, transportation and tourism, federal and municipal services, agriculture and food, and national monuments and icons.

Cyber-physical systems are an important area of emerging national and global security and economic competitiveness. Similar to the need of the United States' computing industry to collaborate in the 1980s to enhance semiconductor manufacturing efficiency (fulfilled by Sematech in Austin, Texas), there now exists a need for collaboration and targeted research and development at the intersection of digital communications and control systems related to critical infrastructure protection and cooperative, multi-institutional cyber security mitigation and incident response.

In order to protect and sustain the operation of cyber-physical systems, infrastructure owners including corporations, government, military, and non-profit organizations need to: (1) enable cyber-physical systems security and cooperative incident management strategies in multi-sector environments; (2) promote public–private partnerships for policy and programs related to the development of human and intellectual capital for education, workforce, economic development systems relevant to cyber-physical systems; (3) create collaborative multi-stakeholder laboratories and research, development, and commercialization initiatives to promote public welfare and the security of critical infrastructure systems and processes; and (4) design necessary approaches to science, technology and public policy that effect public safety and cyber-physical systems security both procedurally and systematically.

Acknowledgments This paper would not have been possible without the generous support of Dr. Greg White from the Center for Infrastructure Assurance and Security and the Society for Design and Process Science including Dr. Murat Tanik, Dr. John Carbone, Dr. Abidin Yildirim, and Stan Gatchel. The work to extend process and design science into primary, secondary and post secondary education realms is important and I am grateful for your recognition and encouragement. Diffusion of design and process science into the fabric of society is important to the economic and civil sustainability of the world.

References

1. R. Y. Fahmida, "GM's OnStar, Ford Sync, MP3, Bluetooth Possible Attack Vectors for Cars", e-week.com, IT Security & Network Security News, March 16, 2011 last accessed on July 22, 2012 at http://www.eweek.com/c/a/Security/GMs-OnStar-Ford-Sync-MP3-Bluetooth-Possible-Attack-Vectors-for-Cars-420601/.

2. Al Arabiya with Agencies, "Iran denies role in cyberattacks against Gulf oil and gas companies", Al Arabiya News, Sunday, October 14, 2012 last accessed October 16, 2012 at http://english.alarabiya.net/articles/2012/10/14/243682.html.
3. S. Gorman, "Electricity Grid in U.S. Penetrated By Spies", Wall Street Journal, Technology, April 4, 2009 last accessed on October 17, 2012 at http://online.wsj.com/article/SB123914805204099085.html.
4. Executive Order 13010—Critical Infrastructure Protection, Federal Register, July 17, 1996. Vol. 61, No. 138. in Critical Infrastructure and Key Assets: Definition and Identification, CRS Report for Congress, October 1, 2004 last accessed October 16, 2012 http://www.fas.org/sgp/crs/RL32631.pdf.
5. Sharing Information with the Private Sector, The White House, George W. Bush, n.d. last accessed October 16, 2012 at http://georgewbush-whitehouse.archives.gov/nsc/infosharing/sectionV.html.
6. T. Delbert in J. Brazell, "Internet 2.0: What's Next In Computing—How Cyber-Physical Systems Are Changing Business Commerce, and the Economy", Texas CEO Magazine, April 29, 2012 last accessed on July 22, 2012 at http://texasceomagazine.com/tag/jim-brazell/.
7. Mechatronics homepage, Rensselaer Polytechnic Institute (RPI), n.d. last accessed July 22, 2012 at http://multimechatronics.com/.
8. "Future Work Skills 2020", Apollo Research Institute, April 14, 2011 last accessed July 22, 2012 at http://apolloresearchinstitute.com/research-studies/workforce-preparedness/future-work-skills-2020.
9. J. B. Brazell, L. Donoho, J. Dexheimer, R. Hanneman, G. Langdon, and E. Evans, "M2M: The Wireless Revolution. A Technology Forecast, Implications for Community Colleges in the State of Texas", Texas State Technical College System and IC2 Institute, University of Texas Austin, 2005 last accessed October 17, 2012 at http://forecasting.tstc.edu/wp-content/uploads/2010/12/M2M_The_Wireless_Revolution-TSTC_Forecast-2005.pdf.
10. J. H. Vanston, H. Elliott, J. Irwin, J. Brazell, and M. Bettersworth, "Mechatronics Forecast: Implications for Texas Community and Technical Colleges", Texas State Technical College System, 2007 last accessed October 17, 2012 at http://forecasting.tstc.edu/wp-content/uploads/2010/12/Mechatronics-TSTC_Forecast-2007.pdf.
11. J. Brazell, email interview with Abidin Yildirim, Society for Design and Process Science, July 6, 2012.
12. E. Evans, M. Sekora, A. Cavalli, K. Chan, J. Heo K. Kan, Y. Kuang, P. Mohandas, X. Zhang, and J. Brazell, "Digital Convergence Initiative: Creating Sustainable Competitive Advantage in Texas", Greater Austin-San Antonio Corridor Council, 2005 last accessed October 17, 2012 at http://www.dcitexas.org/DCI_report.pdf.
13. "Converging Technologies for Improving Human Performance—Nanotechnology, Biotechnology, Information Technology, and Cognitive Science", edited by M. C. Roco and W. S. Bainbridge, National Science Foundation (NSF) and Department of Commerce (DOC), 2002 last accessed on July 22, 2012 at http://www.wtec.org/ConvergingTechnologies/Report/NBIC_pre_publication.pdf.
14. Executive Office of the President President's Council of Advisors on Science and Technology, "Report to the President and Congress, Designing a Digital Future: Federally Funded Research and Development in Network and Information Technology", December 2010, Last accessed on October 17, 2012 at http://www.whitehouse.gov/sites/default/files/microsites/ostp/pcast-nitrd-report-2010.pdf.
15. Occupational employment projections to 2018, Monthly Labor Review, U.S. Bureau of Labor Statistics, November 2009 in "Report to the President and Congress, Designing a Digital Future: Federally Funded Research and Development in Network and Information Technology", December 2010 last accessed on October 17, 2012 at http://www.whitehouse.gov/sites/default/files/microsites/ostp/pcast-nitrd-report-2010.pdf.
16. J. B. Brazell, email interview with G. White, Center for Infrastructure Assurance and Security, University of Texas San Antonio, October 17, 2012.

Chapter 3
A Regional and Transdisciplinary Approach to Educating Secondary and College Students in Cyber-Physical Systems

Cliff Zintgraff, Carolyn Wilson Green and John N. Carbone

Background

In the U. S. President's Council of Economic Advisors (PCAST) report titled *Leadership Under Challenge: Information Technology R&D in a Competitive World* [1], the Council notes four research areas which "should receive disproportionately larger increases because they address issues for which progress will have both the greatest effect on important applications and the highest leverage in advancing NIT capabilities." The first area listed is "NIT Systems Connected with the Physical World … cyber-physical systems" (p. 2). Cyber-physical systems will be an arena of intense economic competition, with the PCAST report noting EU investment of over US$7 billion between 2007 and 2013. The U. S. National Research Council (2001) [2] reports "the verge of another revolution…networked systems of embedded computers…could well dwarf previous milestones in the information revolution" (pp. 1–2).

Defined by the PCAST report as "NIT systems connected with the physical world" (p. 31), cyber-physical systems are often real-time and critical in nature. They include defense and intelligence systems, air traffic control, power grids, water supply systems, vehicles, clinical and home healthcare, environmental monitoring, industrial process control, and ground transportation management. Cyber-physical systems "[synthesize] knowledge from the physical sciences, mathematics, engineering, biological sciences, computer science, and other fields to model and simulate such systems in their full complexity and dynamics; [this includes] the interactions among potentially many dynamic systems and

C. Zintgraff (✉)
University of North Texas, Denton, USA
e-mail: cliffzintgraff@my.unt.edu

C. W. Green
Texas A&M University-San Antonio, San Antonio, USA

J. N. Carbone
Raytheon Company, Garland, USA

S. C. Suh et al. (eds.), *Applied Cyber-Physical Systems*,
DOI: 10.1007/978-1-4614-7336-7_3,
© Springer Science+Business Media New York 2014

15

components in uncertain environments" (p. 32). Lee [3] associates cyber-physical systems with "high confidence medical devices and systems, traffic control and safety, advanced automotive systems, process control, energy conservation, environmental control, avionics, instrumentation, critical infrastructure control (electric power, water resources, and communications systems for example), distributed robotics (telepresence, telemedicine), defense systems, manufacturing, and smart structures" (pp. 1–2) and notes that "the resulting research agenda is enormous, spanning most disciplines within computer science and several within other engineering disciplines" (p. 8). The PCAST report says that we must "enable industry and universities to collaborate on pre-competitive research in these systems" (p. 33).

To address workforce demand for cyber-physical systems, industry and universities will need a qualified workforce. The complexity of these systems strongly suggests that workers and researchers in these fields cannot be narrowly focused. They need to be comfortable working with people and concepts from disciplines besides their own. Ideally, their training will be from more than one discipline.

Literature Review

The authors propose a regional, transdisciplinary approach to cyber-physical systems education. The literature foundation for this proposal includes *transdisciplinarity*, *workforce context*, the *centrality of regional approaches*, and *lessons drawn from related industry clusters*.

The Transdisciplinary Approach

This chapter views the challenge of educating secondary and college students in cyber-physical systems through the lens of *transdisciplinarity*. Many find the concept difficult to distinguish from other multi-discipline approaches. Various overlapping definitions and distinctions between related multi-discipline approaches are found in the literature. As an essential foundation of this paper, it is important to identify key transdisciplinary concepts that are of particular relevance for cyber-physical systems education, in order to highlight their special nature and implications, and to clarify the meaning of the term as applied in this paper. The reader is encouraged to focus on the principles essential to transdisciplinarity, rather than on the finer points of multi-discipline definitions.

Ertas (2000) [4], Tate et al. [5], and de Freitas et al. [6] provide varying interpretations of transdisciplinarity and related concepts. Problem-solving methods involving insights from multiple disciplines are variously referred to as *multi*disciplinary, *inter*disciplinary, *cross*disciplinary and *trans*disciplinary. Definitions of these terms often overlap but tend to differ in important ways.

Approaching these distinctions narratively, it is possible to approach a problem by: (1) applying a single discipline's concepts or methods to the question at hand; (2) applying multiple disciplines to the problem, using each independently; (3) applying multiple disciplines to the problem, allowing the disciplines to co-mingle, potentially creating new concepts and methods; or (4) bringing multiple disciplines to bear, refusing to be bound by any of the disciplines' rules and methods and aggressively mixing ideas from within and outside the disciplines, unafraid to pick a concept or method from anywhere, interact with anyone, or create new concepts and methods at will.

Ertas (2000) [4] described the goals of a transdisciplinary education and research endeavor. In the process, he describes the nature of the transdisciplinary model, suggesting it "transcends the artificial boundaries imposed by traditional academic organizational structures and directly addresses ... [problems] related to the solution of large and complex problems by teams consisting of many people from diverse backgrounds" (p. 15). Madni [7] provides further insight into the process of developing a transdisciplinary perspective, noting that "[bridging] independent disciplines typically requires extending them, reconciling their differences, and unifying the knowledge associated with them in new and novel ways" (p. 1).

Figure 3.1 provides a summary of these concepts and an illustration of their distinct approaches. The non-partisan "unicorns" represent our view of transdisciplinarity in this paper. Calling on definitions and elaboration from Tate et al. [5] and de Freitas et al. [6], we highlight these transdisciplinary principles as foundational to the approach of this paper:

- Disciplines are good. They provide philosophy, concepts and methodologies for approaching problems that once were beyond human capacity to address.
- Like any good thing, disciplines have negative impacts when they lead to mental rigidity in situations novel to that discipline.
- Complex problems are multi-dimensional. They are scientifically complex; operate inside complex systems; involve hard and so-called soft sciences; and at a minimum appear unpredictable because of their complexity.

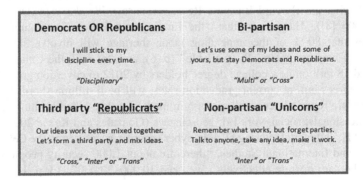

Fig. 3.1 Distinguishing *Trans*disciplinarity

- Problems like these should be approached by collaborators from every relevant discipline, within practical reason.
- Collaboration levels are high, and collaborators set aside disciplinary restrictions. Openness and a high level of mutual respect are cultural values.
- An iterative think-and-do approach is warranted, valuing philosophical inquiry, theoretical inquiry, development of solutions, and critical analysis and iteration.
- The possibility exists that a new discipline is being created with its own philosophy, concepts and methods. A transdiscipline may (or may not) be a stepping stone to a new discipline.
- The fundamental enabler for transdisciplinarity is creating the environment that allows all participants to act according the principles above.

The authors view the essential task of transdisciplinary inquiry as creating an environment where disciplinarians can, speaking colloquially, *let their hair down, relax, and create new solutions to problems*. This involves characterizing the problem domain, characterizing the larger system, identifying the relevant disciplines, and creating a platform for common work. The remainder of this paper considers cyber-physical education in the terms of this section.

Workforce Concerns: Meta-Context and Implications

The challenge of fostering industry and university collaboration raises the question of who will work in these companies and colleges to accomplish the important work at hand. This is not an idle question given the numerous examples where regions and industry clusters have struggled to reach their full potential for lack of a workforce and regional support ecosystem. A quick review of literature in other fields reveals a sobering meta-context. In the U. S., workforce shortages are acute in numerous fields. There are multiple factors causing these shortages, including the aging of the Baby Boom generation, shifting demographics, and the decline of selected industries and the rise of others; this last factor is a significant underlying cause of what the Harvard Business Review calls a *skills shortage* (Dychtwald et al. 2006 [8]). They report that "the Employment Policy Foundation (EPF) estimates that 80 % of the impending labor shortage will involve skills, not numbers of workers potentially available" (p. 6); and "overall, the United States will need 18 million new college degree holders by 2012 to cover job growth and replace retirees but, at current graduation rates, will be 6 million short" (p. 11). Shortages are reported across all sectors: in medical fields, allied health, geriatric care professionals, social work [9]; geosciences (Gonzales and Keane 2006 [10]); energy [11]; and manufacturing [12]. In cyber security, according to the Center for Strategic and International Studies, "there are about 1,000 security people in the U. S. [who can] operate effectively...we need 10,000–30,000" [13, p. 1].

Workforce challenges are not limited to the U. S. In India, where workers of all types (unskilled, skilled and professional) had to be imported from China to build the Delhi airport, "the country needs to train 500 million skilled laborers by 2022 if its current economic growth is to continue" [14]. The health workforce crisis is global, affecting the U. S., Europe and Africa; [15, 16] the utility infrastructure workforce faces challenges in North America, Europe and Japan [17]; China faces "structural problems [with] a labor shortage in some professions, industries and regions;" South China's Guangdong province, the nation's production and export base, is short "about 800,000 laborers this year" [18].

This meta-context informs us that cyber-physical systems as a global field will demand systematic efforts to raise a competent, scalable workforce. The same underlying challenges from other fields apply. The multiple disciplines involved and the associated concerns of workforce education and training, college experiences and K-12 preparation complicate matters. Poovendran (2010) [19] asserts that "current educational and workforce training frameworks are not sufficient" (p. 1365), citing the need for courses that integrate across computer science, engineering, anthropology and sociology. The transformation needed moves beyond interaction among existing disciplines. We must "develop a workforce with new skills…where the traditional disciplinary boundaries need to be redefined in all levels of training and education" (Dumitrache 2010 [20], p. 4).

Regional Motivators: Education, Training, Workforce, and Economic Development

While there are clear global, national, regional and organizational layers to consider, much focus on education, training, innovation and economic development occurs at regional, i.e., city or greater metropolitan area scope. Cooke [21] defines a region as being homogenous around specific criteria, possessing internal cohesion, and defined and driven by its internal cohesion needs, vs. being driven from size considerations or political borders. Cooke suggests that the [22] seminal definition of an *industry cluster* can help define a region. Porter defines an industry cluster as "a geographically proximate group of interconnected companies and associated institutions in a particular field, linked by commonalities and complementarities" (p. 1998). Mills et al. [23] note characteristic attributes: participants in the cluster are concentrated, economically motivated, transact intensively, and cover a broad range of business activities like R&D, marketing, sales, services, finance, and operations.

Porter also notes the importance of regional support systems, an attribute of clusters that receives additional focus in the *technopolis* work of Smilor et al. [24]. In their *technopolis wheel* model, the authors note the broad array of elements needed to support a technopolis. The term *polis* evokes the "city-state, reflect[ing] the balance between the public and private sectors" (p. 50). One factor especially important in a technopolis is "attraction of major technology companies" (p. 50). Mills et al. [23]

note that an important factor to companies is "related research, education, and training institutions such as universities, community colleges, and workforce training programs" (p. 9), reinforcing the assertions of the technopolis model.

The regional opportunities afforded by cyber-physical systems powerfully align with classic regional interests: education, economics, quality of life, and both practical and noble desires to contribute to important national and global challenges. Kitson et al. [25] report "there is now widespread agreement that we are witnessing the 'resurgence' of regions as key loci in the organization and governance of economic growth and wealth creation" (p. 991). In Mills et al. [23], whose writings appeared in the Brookings Institute publication *Blueprint for American Prosperity: Unleashing the Potential for a Metropolitan Nation*, the authors make the case for regional strategies. They note the relative inactivity of the federal government's support for region-based strategies, despite the prominence of these strategies in the foreground of overall economic development discussion, and they advocate more U.S. federal support. Cooke (2001) [26] explains the drivers of the trend toward region-based initiatives, referring to the region as the "more natural economic zone" (p. 4). With globalization as the macro trend, Cooke views the nation-state's relative strategic position vs. regions as declining. When regions have developed robust clusters of core actors and supporting organizations, such economic communities of interest are highly motivated to compete, and they interact directly with domestic and foreign partners, directly attracting foreign direct investment. Globalization drives foreign direct investment as economic actors seek the most efficient location for business activities. A virtuous cycle ensues, as "the to and fro of adjustment between companies, markets, public authorities, research institutes, training institutions and social partners [transform] each, while creating the elements of an innovative framework that may encompass and stabilise them all" (p. 4).

It is important to recognize the deeper drivers for regional strategies. Kitson et al. [25] believe that regional strategies are more about "long-run prosperity than...restrictive notions of competing over shares of markets and resources...ultimately competitive regions and cities are places where both companies and people want to locate and invest" (p. 997). K-12 education, higher education, workforce training, robust industry clusters, and high-wage jobs, not to mention arts and culture, all connect to long-term prosperity and quality of life for citizens.

Drawing Lessons from Related Clusters

The authors performed a search for formal industry clusters in the field of cyber-physical systems.[1] No formal clusters were found. The authors speculate this is

---

[1] Some sites self-identified as clusters. However, the authors chose to require clear documentation of specific industry connections and evidence of a local collection of interdependent industries. Supporting data were not immediately forthcoming for the sites investigated.

simply for lack of formal definition, as based on a simple web search and the personal knowledge of the authors, there are U.S.-based clusters of *activity* in Austin, Texas; Dallas, Texas; and around Oregon State University [27] and the University of Connecticut [28]. Nevertheless, related clusters in cyber security offer a robust proxy from which lessons can be drawn for clusters in cyber-physical systems. Noted are the similarities in secondary school preparation, college coursework, job attributes, and field complexity, not to mention the concrete overlap between the two fields. U.S.-based cyber security clusters identified are geographically based in Maryland; San Diego, California; and San Antonio, Texas.

The *Maryland Cybersecurity Center* (*MC²*) at the University of Maryland "[partners] with government and industry to provide educational programs to prepare the future cybersecurity workforce, and develop new, innovative technologies to defend against cybersecurity attacks." MC^2's approach is holistic, interdisciplinary, and includes solutions engineering, computer science, economics, social sciences and public policy. Cybersecurity education is workforce-relevant, applying a "teaching hospital" model that includes "undergraduates, graduate students and industry professionals" working "practical cybersecurity challenges" (MC^2[29–33], About the Maryland Cybersecurity Center). MC^2 explicitly promotes technology commercialization and entrepreneurship through connections to various university programs and competitions. Industry partners include Northrop–Grumman, Lockheed-Martin, SAIC, Google, Sourcefire and CyberPoint (MC^2[29–33], UMD and CyberPoint announce new cybersecurity partnership); (MC^2 [29–33], Corporate partnerships in cybersecurity). A Cybersecurity Club has over 300 student members engaging in competitions, and hearing from distinguished speakers and industry experts (MC^2 [29–33], Cybersecurity education). There are cybersecurity concentrations, and their ACES program is claimed as "the nation's first undergraduate honors program in cybersecurity" (MC^2 [29–33] Advanced cybersecurity experience for students). In addition to MC^2, a *National Cybersecurity Center of Excellence* (*NCCOE*) will be operated by the National Institute of Standards and Technology (NIST), the State of Maryland, and Montgomery County, Maryland. NCCOE will include or connect to efforts in workforce development; K-12 STEM education; federal, state and local program integration; and connection to U. S. Cyber Command at Ft. Meade (NIST Tech Beat, [34]).

Based in San Diego, California, *Securing our eCity*[®] (*SOeC*) is a regional collaboration with the mission "to enable every San Diegan to live, work and play safely in the cyber world" (SOeC [35–37], Model city). SOeC began as an effort to educate San Diego-area citizens and businesses to adopt smart cyber practices and protect themselves online. The effort is a broad partnership, with key stakeholders including ESET Internet Security, the San Diego Regional Chamber of Commerce, San Diego Business Journal, San Diego Gas and Electric, San Diego State University (SDSU), and the University of California at San Diego (UC San Diego). An additional 134 organizations representing a diverse set of large and small businesses; military, government organizations and community organizations; and professional associations are listed as stakeholders. SOeC's activities include free workshops for businesses; a collection of educational resources for persons, families and businesses; references to local Science, Technology, Engineering and Math (STEM) events; and a yearly Cybersecurity Symposium and Awards event. A major effort is the San Diego Mayors' Cyber Cup Challenge, now in its fourth year as a "cyber defense competition [that] encourages mentoring of middle and high school students through formation of local school teams that compete in a realistic cybersecurity exercise, in which they detect and defend computer systems against hackers intent on stealing data" (SOeC [35–37], Sneak peak). In 2010, SOeC was awarded the Best Local/Community Cyber Security Challenge award from The Department of Homeland Security.

In San Antonio, Texas, the *Cyber City, USA* initiative is a broad-based community collaboration promoting education, business, and workforce development in the cyber security field. San Antonio's effort is anchored by the city's large military presence, which includes "the nation's largest military installation...Joint Base San Antonio, with nearly 80,000 Department of Defense personnel assigned" (San Antonio Greater Chamber of Commerce, [38]) and the 24th Air Force, which "establishes, operates, maintains and defends Air Force networks" ([39], 24th Air Force fact sheet). Military missions are supported by numerous large defense contractors, small companies, and information security startup companies. Fortune 500 companies support the effort, including USAA and Valero Energy Corporation, and significant support is provided by Rackspace. Three local universities are designated as centers of excellence in information security by the National Security Agency and Department of Homeland Security, including the University of Texas at San Antonio (UTSA), Our Lady of the Lake University, and Texas A&M University-San Antonio. The Cyber Innovation and Research Consortium (CIRC) was formed in 2007 to "[link] San Antonio area academic institutions" and "[work] closely with government, industry, workforce, and economic development initiatives to form a robust public–private partnership" (San Antonio Greater Chamber of Commerce, [40]). The CIRC includes five local universities and the local community college system. The CIRC is expanding to include local high school programs. The Information Technology and Security Academy (ITSA) is a high school dual credit program operated since 2002 by Alamo Colleges, the local community college system, with credit articulated into degree programs at Alamo Colleges, Texas A&M University-San Antonio, UTSA, Our Lady of the Lake University and St. Mary's University [41]. The community has organized to compete in the national Cyber Patriot Cyber Defense Competition, entering 54 teams for competition in 2013, awarding the Cyber Patriot Mayors' Cup, and fielding the national competition's winning team, from ITSA. The City of San Antonio has been named a Cyber Patriot Center of Excellence. The UTSA Center for Infrastructure Assurance and Security (CIAS) is specifically concerned with the security of cyber-physical infrastructure. CIAS has developed a Community Cyber Security Maturity Model (CCSMM) as the basis for "exercises, seminars and workshops" conducted at the community and state level ([42], CIAS). The development of a community-wide Program of Study clearinghouse for educators and students covering the cyber security and cyber-physical systems programs in this region is the topic of the final section of this paper.

Considering these clusters through the lens of the technopolis model highlights the importance of robust, regionally based ecosystems in support of clusters. The technopolis model consists of seven main elements: universities, large corporations, emerging companies, federal government, state government (including education), local government, and support groups (chambers, business associations, community advocacy groups). The University of Maryland, UC San Diego, SDSU, UTSA and Texas A&M University-San Antonio play anchor roles in their communities. Northrop–Grumman, Lockheed-Martin, SAIC, Google, ESET Internet Security, USAA, Valero and Rackspace provide key corporate support. Emerging startups are mentioned prominently on the web sites of these programs. The federal government plays a crucial role, including the support of NSA and DHS and the prominent role of military organizations in cyber security. State governments fund the universities and assist with education and workforce policies that support technology-driven economic development and education. State or local governments fund community colleges that are at the forefront of building a skilled technology workforce. Local governments create education and business friendly policies and bring focus to local industries. Support groups like chambers of commerce and advocacy groups provide relatively turf-free settings that facilitate collective impact. The examples of these cyber security industry clusters, viewed through the lens of the technopolis model, demonstrate the importance of regionally-based strategies that set vision and coordinate resources to achieve maximum community impact.

Forming the Transdiscipline

As noted by Poovendran (2010) [19] and Dumitrache (2010) [20], addressing the workforce challenge associated with the burgeoning world of cyber-physical systems will require new educational and workforce frameworks. Transdisciplinary education programs will be needed that address the complexity and dynamics of networking and information technology systems connected with the physical world [1]. A key ingredient will be the development of a provisional definition of the *cyber-physical systems education transdiscipline*. Such a definition can be the starting point for advancing research and defining a body of knowledge for preparation of professionals who conceive, design, develop, deploy, defend and support cyber-physical systems. Cyber-physical workforce development will require identifying potential job categories associated with the system life cycle and defining the kinds of work these professionals might do, the types of knowledge and skills they need to possess, and the types of academic programs needed to help prepare them to work in this field. Academic program development would draw on these foundations to identify program types and levels (e.g., high school-, associate-, baccalaureate-, masters- and/or doctoral-level programs) and instructional methodologies appropriate to preparing students for these careers. Highly related is the need to establish strong partnerships among academic, industry, government and community leaders to garner input and support and to facilitate communication about local workforce needs and career opportunities afforded by academic programs associated with cyber-physical systems. To be effective, these activities must proceed in a coordinated manner. To illustrate how this might be done, the authors will explore possible approaches to developing a provisional cyber-physical education transdiscipline, establishing an initial set of job categories with their knowledge and skill requirements, and identifying a framework for education and workforce development activities.

Conversations about what constitutes the *cyber-physical systems transdiscipline* have already begun. Lee [3, 43, 44] has suggested that the disciplines necessary to advance research and development of education programs include computer science, mechanical and electrical engineering, mathematics, physics, and system theory, noting that the intellectual core of cyber-physical systems is modeling and meta-modeling (2010). Pappas [45, 46] also emphasizes the importance of modeling, mathematics, networking and physics, and adds disciplines associated with high confidence, system verification, certification, robustness, resilience, security, privacy, and trustworthy systems. Poovendran (2010) 19 suggests that sociology and anthropology may provide valuable insights. The requirements for secure, safe, and reliable systems also suggest that risk management, safety engineering and secure system development methods provide additional insights for cyber-physical system development and deployment. Although it is likely that additional disciplines will be identified, the current discussion is a starting point for ongoing elaboration.

As Ertas (2000) [4] noted in his article introducing the Academy of Transdisciplinary Education and Research (ACTER), the academy's education and research objectives also include "the development of new models of learning and innovative teaching environments to complement transdisciplinary curricula. Such innovations will include active learning environments, less dependence on formal lectures, project based learning, learning through teamwork, and motivations that inspire lifelong learning" (p. 16). Teaching and learning methods that share these characteristics include constructionist and constructivist project-based learning which involve "learning by making" and the construction of models or "public entities" as a means of constructing knowledge [47, 48]. Constructivism includes performing authentic tasks in a real-world context, using project-based or case-based learning, reflection, and collaborative construction of knowledge, working in groups in cooperative environments [49]. In addition to learning experiences associated with formal curricula, active learning experiences may also include internships, mentoring, service learning in students' career fields, and participation in career-related competitions. Methods like these can serve as a reference point in the development of teaching and learning models that complement the emerging cyber-physical transdiscipline.

In the transdisciplinary spirit, it is appropriate to note that while less dependence on formal lectures and more on interactive learning environments is appropriate, the most important design consideration is to apply learning theories appropriate to the task at hand. The authors hold the general belief that typical teaching and learning practices rely too heavily on behaviorist so-called *sit-and-get* methods. Nevertheless, various theoretical perspectives should be considered, deployed where appropriate, and integrated in a transdisciplinary fashion. In *Behaviorism, Cognitivism, Constructivism: Comparing Critical Features from an Instructional Design Perspective*, Ertmer and Newby (1993) 50 assert that "many designers are operating under the constraints of a limited theoretical background" (p. 50). Those authors quote Snelbecker [51] in noting that "individuals addressing practical learning problems cannot afford the 'luxury of restricting themselves to only one theoretical position'" (p. 52; [52], p. 8). Ertmer and Newby, citing a turn-of-phrase from Snelbecker, refer to "instructional cherry-picking [that] has been called 'systematic eclecticism'" (1993, p. 70) and argue there has been significant support in instructional design literature for such a position [51]. A mix of methods can be persuasively argued. Nevertheless, it must be remembered that constructivist, constructionist approaches are those the field must move toward. Tate et al. [5] describe the direction of movement from "authority and regurgitation [to] instruction [to] primarily based on facilitation" (p. 13). In this paper's transdiscipline, an opportunity exists to promote this approach in a consistent and coordinated fashion across all grade levels of instruction.

Addressing workforce needs and the educational programs necessary for preparing students entails identifying what types of jobs will need to be filled and the knowledge and skills required. One model to consider is the NIST [53] NICE National Cybersecurity Workforce Framework. The framework identifies seven categories of job specialties associated with cyber security, including *securely provision*, which involves conceptualizing, designing and building secure IT systems; *operate and maintain*, which involves providing system support, administration and maintenance; and *protect and defend*, which involves identification, analysis and mitigation of threats. The framework identifies sample tasks associated with the job category and required knowledge and skills. The framework's job categories range from specialties that would require advanced education at the baccalaureate and graduate levels (e.g., in the *securely provision* category) to specialties that could be potential jobs for students completing associate degrees or high school advanced technology programs (e.g., in the *operate and maintain* category). Given its overlap with many of the types of jobs that would be associated with cyber-physical systems, a similar framework could be an initial model for defining job categories for cyber-physical workforce development. The framework also includes a four-phased call to action which addresses activities that would be necessary to make effective use of the framework: collect and analyze workforce and training data; recruit and retain; educate, train and develop; and engage and energize the workforce and public.

One popular model used for describing and assessing the state of workforce development in particular careers is the *Program of Study*. A Program of Study (POS) documents available academic and training programs that prepare workers for a career field, support services that assist those who are interested in pursuing those programs, and public policy and community support resources that facilitate the success of the workforce development efforts. It is often used to document career preparation involving a sequence of academic of programs (e.g., from baccalaureate to graduate or associate to baccalaureate program) that permits a student to continue to more advanced education while having the option to enter the workforce at different points in the program's sequence. A cyber security POS might, for example, document a path that starts with completion of a high school diploma with an information technology certification, continue to an associate degree in networking and security with additional certifications, followed by completion of a baccalaureate degree in information technology and cyber security. At each point where an academic program is completed, the POS would document career-related job opportunities available upon degree completion.

The OVAE (Office of Vocational and Adult Education) [54] POS design framework, for example, identifies ten components that support the development and effectiveness of a program of study. The framework provides a template for documenting legislation and policies that affect the POS, partnerships that support the program, professional development programs for faculty and administrators associated with the program, accountability and evaluation systems that assess the program's effectiveness, college and career readiness standards associated with the program, POS course sequences, credit transfer agreements, guidance counseling and advisement services, the program's teaching and learning strategies, and the program's student learning outcome evaluation process. The POS framework can be used for planning curriculum and related structures, and for documenting the current state of the POS and planning for improving its quality and effectiveness. It can also be used as a platform for communication and coordination of a community's efforts to advance their workforce development objectives in a particular career field.

Pilot Case Study: Applying the Transdiscipline to San Antonio's Cyber Security and Emerging Cyber-Physical Systems Clusters

In the technopolis model, one challenge is creating and articulating a framework that is strategically compelling, while simultaneously providing opportunities for contribution by a diverse set of partners, without losing collective impact. The authors of this paper are active in the San Antonio education, cyber security and emerging cyber-physical systems communities. They have proposed an education-workforce focused strategy with the potential to engage partners from all seven technopolis elements while addressing key observed challenges. These challenges were observed:

• Cyber, cyber security and information technology of all types is growing rapidly in the community [55].

- Organizations verbally report not being able to fill open positions.
- Educators, parents and students do not demonstrate awareness of local opportunities to complete cyber education pathways and enter the workforce in skilled or professional positions.
- Students are frequently not making the coursework choices required in high school to be accepted into college programs that lead to cyber degrees and jobs.

In considering these challenges through the lens of this paper's transdiscipline, the authors recall the concept of *Programs of Study* and offer it as the core element of a cyber-physical systems education transdiscipline and regional development strategy. The strategy proposed recommends a *regional Programs of Study clearinghouse* as a pragmatic foundation for numerous activities.

The POS clearinghouse approach offers multiple potential benefits: (1) it facilitates education of students and parents about the importance of rigorous STEM high school coursework starting in middle school (grades 6–8); (2) it informs educators and other stakeholders about these same education and career opportunities; (3) by promoting informed secondary school course selection, it reduces the need for remedial education for students entering college; (4) it helps keep college costs in check by helping students and parents clarify their interests, set their goals, and focus on goal-related coursework; (5) it creates POS-wide context within which philosophical and pedagogical approaches can be promoted, crossing traditional boundaries, to strive for improved educational outcomes. In short, a Programs of Study clearinghouse strategy creates clarity and pragmatic context. It is proposed that a Programs of Study clearinghouse will provide a common, robust, and flexible platform, building on the partnership-driven culture of the community, which in turn builds organizational capacity to counsel and educate students, parents, and community stakeholders about the cyber opportunities that exist in their community.

From a systems perspective, the effort to create a Programs of Study clearinghouse will lead to identification of gaps. Funding proposals can be developed to enhance current programs, build new programs, and/or build additional systemic capacity. Awareness of important local programs will facilitate engagement of all stakeholders to support the formal and informal educational activities that are documented in the Programs of Study. The proposed approach also supports building both the skilled and professional workforce, with multiple entry and exit points that provide maximum flexibility to lower income and middle class families.

Pilot Project

Sponsored by the regional Cyber Innovation and Research Consortium (CIRC), a pilot project was engaged to document the cyber pathways for three schools: Texas A&M University-San Antonio, Our Lady of the Lake University, and San Antonio College, a community college campus of the larger Alamo Colleges system.

The goal of the pilot project was to document all cyber-related pathways, and link those pathways in a multiple-entry, multiple-exit format that shows students the various ways they can reach cyber degrees. A second goal of the pilot project was initial connection to cyber-specific high school programs and middle school career and technology education (CTE) courses. An information technology platform called SooperMinds was selected for this project. The platform contains features supporting entry of courses, collection and sequencing of courses toward an end goal, integration of CTE and academic courses, inclusion of extracurricular activities to engage students, and the ability to document multiple, alternative articulations from specific high schools/programs, into local colleges, and on to advanced degrees and the workforce. SooperMinds' features also support communication between students taking courses in a sequence, and the educators operating later programs to which students might aspire (for example, a high school freshman can communicate via SooperMinds with the director of a college program, moderated by their high school teacher). Finally, features allow students and parents to co-explore courses and sequences, and share their preferences among the co-explorers.

During the two-year pilot project, 22 *pathways* were populated into Sooper-Minds and linked together in Programs of Study that cross secondary and college boundaries. A SooperMinds pathway is one segment of a POS; for example, the four years of a high school program, or the two years of a community college associate degree program constitute a pathway. Each pathway is maintained by its host school. An owner of a pathway can initiate a link to other pathways, which can in turn be accepted or rejected by the "target" pathway's owner. With these features, the community is encouraged to self-organize, assume responsibility for content, and collaborate with partner institutions.

Initial Usage and Findings

As noted, three colleges took part in the pilot study, taking responsibility for entry of their colleges' pathways. As part of a larger project that reached 587 middle school students across Texas, seventeen additional middle schools in three local school districts were also exposed to Programs of Study in their local regions. The results from this initial study were sobering, with a small mix of good news for the cyber, cyber security and cyber-physical systems communities. Statewide, the three pathways that received greatest interest from students were Culinary Arts, Fashion Design and Criminal Justice/Law Enforcement. This result reinforces popular conceptions and qualitative results that indicate failure to educate students about cyber, STEM, and other high-paying technology-based education and career opportunities. Nevertheless, the pathways to receive the most interest in the San Antonio region were Bachelor of Applied Arts and Sciences in Information Technology and Security (a Texas A&M University-San Antonio program) and Web Programming Level 1. This provides some early indicators that students can be made aware of programs which offer greater academic challenge and education/career opportunities in their local community.

Ongoing Efforts

CIRC has completed the pilot project and is moving forward with full implementation, proceeding to document pathways for all CIRC partner higher education institutions. The Programs of Study clearinghouse strategy will be used to highlight the breadth of cyber education and jobs in the region for all stakeholders; drive awareness and activity in creating cross-boundary Programs of Study; drive awareness activities for the primary feeder schools for CIRC institutions; provide a view of gaps that require institutional and systemic action; provide a basis for pedagogical discussions across boundaries to pursue improved outcomes; and provide the basis for additional research and programs amenable to external funding. Because Programs of Study are fundamentally multi-disciplinary, they provide context for multi-stakeholder discussions. While such discussions are not required to be transdisciplinary in nature as defined in this paper, the region intends to make them so. The transdisciplinary approach which honors disciplines, while working across and outside disciplines in an integrated and holistic fashion, provides a defining culture for discussions. In the view of the authors, the approach leads to solutions that are not falsely bounded by disciplinary constraints, but are instead grounded in the multi- and extra-disciplinary realities of building a robust education and workforce system for cyber, cyber security, and cyber-physical systems in the region.

Conclusion

This chapter proposed a regionally focused, transdisciplinary approach to creating education-workforce systems that support industry clusters important to local communities. The discussion has been placed in the context of the growth of cyber-physical systems clusters, drawing heavily from the overlapping and somewhat more mature field of cyber security. The transdiscipline proposed creates a holistic, integrated culture, and the pragmatic foundation of Programs of Study provides a basis and context for immediate action and further research.

This study is limited in several ways. It has primarily focused on regions in the United States, and may be less applicable to concerns in other countries. The relative newness of cyber-physical systems required accessing information and lessons from the cyber security field, which may or may not be fully applicable. The Programs of Study clearinghouse strategy applied in the pilot study is in its early stages. Additional potential research questions include: How applicable are these concepts globally? Can effective discourse take place in a complex transdisciplinary context? Can a distributed Program of Study definition strategy be effective? Does promotion of local Programs of Study to middle and high school students positively impact student behavior? Does the context provided by Programs of Study lead to gap identification and actions to address those gaps? Does a Programs of Study strategy lead effectively to fundable projects? The authors propose these as important questions in a global economy where regions must compete to serve the education and workforce interests of their citizens.

The authors wish to acknowledge the IC^2 Institute at the University of Texas at Austin, San Antonio community volunteers Jim Brazell, Joe Sanchez and Chris Cook, Numedeon, Inc., Power Across Texas, the Texas Workforce Commission, and Next Generation Learning Challenges for contributions of ideas and resources that made this effort possible.

References

1. Marburger, J. H., Kvamme, E. F., Scalise, G., & Reed, D. A. (2007). Leadership under challenge: Information technology R&D in a competitive world. An assessment of the federal networking and information technology R&D program. Washington D.C.: Executive Office of the President, President's Council of Advisors on Science and Technology.
2. U. S. National Research Council (2001). Embedded, everywhere: A research agenda for networked systems of embedded computers. Washington, DC: National Academies Press.
3. Lee, E. A. (2006) Cyber-Physical Systems – Are computing foundations adequate? NSF workshop on cyber-physical Systems: Research motivation, techniques and roadmap, October 16-17, Austin, TX. Retrieved on October 14, 2012 at http://ptolemy.eecs.berkeley.edu/publications/papers/06/CPSPositionPaper/Lee_CPS_PositionPaper.pdf.
4. Ertas, A. (2000). The academy of transdisciplinary education and research (ACTER). Journal of integrated design and process science, 4(4), 13-19.
5. Tate, D., Ertas, A., Tanik, M. M., & Maxwell, T., T. (2006). Foundations for a transdisciplinary approach to engineering systems research based on design and process. TheATLAS Module Series: Transdisciplinary Engineering & Science, 2(1), 4–37.
6. de Freitas, L., Morin, E., & Nicolescu, B. (1994). Charter of transdisciplinarity. In *International Center for Transdisciplinary Research, adopted at the First World Congress of Transdisciplinarity, Convento da Arrábida, Portugal.* Retrieved October 20, 2012 from http://emergingsustainability.org/interdisciplinarity/forum/charter-transdisciplinarity.
7. Madni, A. M. (2007). Transdisciplinarity: Reaching beyond disciplines to find connections. *Transactions of the Society for Design and Process Science*, March 2007, 11(1), 1–11.
8. Dychtwald, K., Erickson, T. J., & Morison, R. (2006). Workforce crisis: How to beat the coming shortage of skills and talent. Harvard Business Press.
9. Bial, M. (2005). Looming workforce shortages in aging services: Getting ready on college campuses. *The Journal of Pastoral Counseling: An Annual: The Westchester County Pre-White House Conference on Aging: Transforming Education and Practice in the 21st Century*, XL, 49–60.
10. Gonzales, L. M., & Keane, C. M. (2009). Who will fill the geoscience workforce supply gap?. Environmental science & technology, 44(2), 550-555.
11. Strack, R., Baier, J., & Fahlander, A. (2008). Managing demographic risk. *Harvard Business Review*, 86(2), 119–128.
12. Morrison, T., Maciejewski, B., Giffi, C., Stover DeRocco, E., McNelly, J., & Carrick, G. (2011). Boiling point? The skills gap in US manufacturing. *Deloitte Consulting & The*

Manufacturing Institute. Retrieved on October 20, 2012 at http://www. themanufacturinginstitute.org/ ~/media/A07730B2A798437D98501E798C2E13AA.ashx.

13. Evans, K. (2010). A human capital crisis in cybersecurity. Washington, DC: Center for Strategic and International Studies.

14. Overdorf, J. (2011). Blame Ghandi: Did the great pacifist kill India, Inc.?. *GlobalPost*. Retrieved on October 20, 2012 from http://www.globalpost.com/dispatch/news/regions/asia-pacific/india/110616/shiva-rules-india-economy-labor-shortage-vocational-training-corporate.

15. Action for Global Health (2011). Addressing the global health workforce crisis: Challenges for France, Germany, Italy, Spain and the UK. Retrieved on October 20, 2012 from http://ec.europa.eu/health/eu_world/docs/ev_20110915_rd02_en.pdf.

16. Conway, M. D., Gupta, S., & Khajavi, K. (2007). Addressing Africa's health workforce crisis. *The McKinsey Quarterly*.

17. Zeiss, G. (2011). Utility workforce crisis in Europe: Participation rates among older workers. Retrieved on September 25, 2012 from http://geospatial.blogs.com/geospatial/2011/07/utility-workforce-crisis-in-europe.html.

18. Yuanyuan, Hu (2012). Workforce shortage a structural problem. China Daily. Retrieved on September 25, 2012 from http://europe.chinadaily.com.cn/china/2012-04/16/content_15053863.htm.

19. Poovendran, R. (2010). Cyber–physical systems: Close encounters between two parallel worlds [Point of view]. Proceedings of the IEEE, 98(8), 1363-1366.

20. Dumitrache, I. (2010). The next generation of Cyber-Physical Systems. Journal of Control Engineering and Applied Informatics, 12(2), 3-4.

21. Cooke, P., & Memedovic, O. (2003). Strategies for regional innovation systems: Learning transfer and applications. United Nations Industrial Development Organization.

22. Porter, M. (1998). On competition. Boston: Harvard Business School Press.

23. Mills, K. G., Reynolds, E. B., & Reamer, A. (2008). Clusters and competitiveness: A new federal role for stimulating regional economies. Washington, DC: Brookings Institution.

24. Smilor, R. W., Gibson, D. V., & Kozmetsky, G. (1989). Creating the technopolis: high-technology development in Austin, Texas. *Journal of Business Venturing*, 4(1), 49–67.

25. Kitson, M., Martin, R., & Tyler, P. (2004). Regional competitiveness: An elusive yet key concept? *Regional Studies*, 38(9), 991–999.

26. Cooke, P. (2001). Strategies for regional innovation systems: learning transfer and applications. United Nations Industrial Development Organization.

27. Oregon State University Mechanical Industrial and Manufacturing Engineering (2012). Complex cyber-physical systems research. Retrieved October 15, 2012 from http://mime.oregonstate.edu/research/clusters/ccps.

28. University of Connecticut School of Engineering (2012). Advanced manufacturing/materials genomics. Retrieved October 15, 2012 from http://www.jobs.uconn.edu/faculty/clusters/advanced_manufacturing_genomics.html.

29. Maryland Cybersecurity Center (2012). About the Maryland Cybersecurity Center. Retrieved October 15, 2012 from http://www.cyber.umd.edu/.

30. Maryland Cybersecurity Center (2012). Advanced cybersecurity experience for students. Retrieved October 15, 2012 from http://www.cyber.umd.edu/documents/2012-ACES-Fact-Sheet-Final-Updated.pdf.

31. Maryland Cybersecurity Center (2012). Corporate partnerships in cybersecurity. Retrieved October 15, 2012 from http://www.cyber.umd.edu/partners/index.html.

32. Maryland Cybersecurity Center (2012). Cybersecurity education. Retrieved October 15, 2012 from http://www.cyber.umd.edu/education/index.html.

33. Maryland Cybersecurity Center (2012). UMD and CyberPoint announce new cybersecurity partnership. Retrieved October 15, 2012 from http://cyber.umd.edu/news/news_story.php?id=6343.

34. National Institute of Standards and Technology (2012). NIST establishes national cybersecurity center of excellence. *NIST Tech Beat*. Retrieved October 18, 2012 from http://www.nist.gov/itl/csd/nccoe-022112.cfm.

35. Securing Our eCity Foundation (2012). About Securing Our eCity®. Retrieved October 18, 2012 from http://securingourecity.org/about.
36. Securing Our eCity Foundation (2012). Sneak peak: San Diego Mayor Jerry Sanders to issue county-wide cyber cup challenge at Securing Our eCity® event (Oct. 11–12). Retrieved October 18, 2012 from http://securingourecity.org/blog/2012/10/11/sneak-peak-san-diego-mayor-jerry-sanders-to-issue-county-wide-cyber-cup-challenge-at-securing-our-ecity-event-oct-11-2/.
37. Securing Our eCity Foundation (2012). Model city. Retrieved October 18, 2012 from http://securingourecity.org/model-city.
38. San Antonio Greater Chamber of Commerce (2012b). CyberCityUSA.org. Retrieved October 20, 2012 from http://cybercityusa.org/education.
39. 24th Air Force (2012). 24th Air Force fact sheet. Retrieved October 19, 2012 from http://www.24af.af.mil/library/factsheets/factsheet.asp?id=15663.
40. San Antonio Greater Chamber of Commerce (2012a). The Cyber Innovation and Research Consortium. Retrieved October 20, 2012 from http://cybercityusa.org.
41. Alamo Colleges (2012). Information Technology & Security Academy. Retrieved October 19, 2012 from http://www.alamo.edu/academies/itsa/.
42. Center for Infrastructure Assurance and Security (2012). CIAS. Retrieved October 19, 2012 from http://cias.utsa.edu/.
43. Lee, E. A. (2008). Cyber-physical systems: design dhallenges, technical report no. UCB/EECS-2008-8. Retrieved on October 14, 2012 at http://chess.eecs.berkeley.edu/pubs/427.html.
44. Lee, E. A. (2009). Introducing Embedded Systems: A Cyber-Physical Approach, Workshop on Embedded Systems Education, Grenoble, France, October 15, 2009. Retrieved on October 14, 2012 at http://ptolemy.eecs.berkeley.edu/publications/papers/06/CPSPositionPaper/Lee_CPS_PositionPaper.pdf.
45. Pappas, G. J. (2009). Cyber-physical Systems: Educational Challenges, CPS Forum, San Francisco, April 15, 2009. Retrieved on October 14, 2012 at http://varma.ece.cmu.edu/CPS-Forum/Presentations/Pappas.pdf.
46. Pappas, G. J. (2012). Cyber-physical systems: Research challenges. NIST foundations for innovation in cyber-physical Systems workshop, March, 2012. Retrieved on October 14, 2012 at http://events.energetics.com/NIST-CPSWorkshop/pdfs/NIST-CPS_Pappas.pdf.
47. Papert, S. (1980). *Mindstorms: Children, computers, and powerful ideas.* New York: Basic Books.
48. Papert, S. and Harel, I. (1991). *Situating Constructionism.* Retrieved on July 28, 2011 at http://www.papert.org/articles/SituatingConstructionism.html.
49. Jonassen, D. H. (1994). Thinking technology: Towards a constructivist design model. *Educational Technology*, 34(4), 34–37.
50. Ertmer, P. & Newby, T. (1993). Behaviorism, cognitivism, constructivism: comparing critical features from an instructional design perspective. Performance Improvement Quarterly, 6(4), 50-72.
51. Snelbecker, G. E. (1983). Learning theory, instructional theory, and psychoeducational design. New York: McGraw-Hill.
52. Snelbecker, G. E. (1989). Contrasting and complementary approaches to instructional design. In CM. Reigeluth (Ed.), *Instructional theories in action* (pp. 321-337). Hillsdale, NJ: Lawrence Erlbaum Associates.
53. NIST (2011). The National Cyber Security Workforce Framework. Retrieved on October 14, 2012 at http://csrc.nist.gov/nice/framework/documents/national_cybersecurity_workforce_framework_printable.pdf.
54. OVAE (2010). Career and Technical Programs of Study: A Design Framework. Retrieved on October 14, 2012 at http://cte.ed.gov/file/POS_Framework_Unpacking_1-20-10.pdf.
55. Butler, R., & Stefl, M. (2009). Information Technology in San Antonio: Economic impact in 2008. Retrieved October 19, 2012 from http://www.sachamber.org/cwt/external/wcpages/wcwebcontent/webcontentpage.aspx?contentid=1367.

Chapter 4
Cyber-Physical Systems and STEM Development: NASA Digital Astronaut Project

U. John Tanik and Selay Arkun-Kocadere

Introduction

Cyber-physical systems can be developed to extend the field of medical informatics with an array of sensors that report data to a device capable of making medical decisions to support clinicians with distributed and embedded artificial intelligence capability. Specifically, a prototype of this system can be built at the university level to teach engineering principles including software engineering in context of cyber-physical system specifications, called the *Health Quest* CDSS. Students taking courses in senior design, as well as graduate students, are uniquely qualified to develop the specifications for a clinical decision support system (CDSS) that has both software and hardware components using languages such as Unified Modeling Language (UML) and Systems Modeling Language (SysML) [1], respectively within a cyber-physical system architectural framework. Education of young students from kindergarten to high school (K-12) can take place in a portal, where the design process is observed and mentoring can take place in the Science, Engineering, Technology, and Mathematics (STEM) component of the project [2]. Furthermore, this CDSS system specification and interactive STEM online portal development functions as an extension to the NASA Digital Astronaut Project [3].

Science, technology, engineering and mathematics (STEM) education is considered critical to society in this technologically demanding age. STEM education is considered as an interdisciplinary bridge among technical disciplines. As a result, STEM education eliminates the barriers among the four disciplines, by

U. John Tanik (✉)
Department of Computer Science, Indiana University-Purdue University,
Fort Wayne, IN, USA
e-mail: jtanik@gmail.com

S. Arkun-Kocadere
Department of Computer Education and Instructional Technologies,
Hacettepe University, Ankara, Turkey

S. C. Suh et al. (eds.), *Applied Cyber-Physical Systems*,
DOI: 10.1007/978-1-4614-7336-7_4,
© Springer Science+Business Media New York 2014

integrating them into a single teaching and learning environment. Therefore, NASA Digital astronaut project is an ideal project to achieve these goals.

This medical guidance system can be built by both graduate and undergraduate students, who can provide guidance/mentorship to K-12 students online throughout all development phases of the *Health Quest* system, which may also include implementation, depending on the budget. The Information Analytics and Visualization (IAV) Center of Excellence at IPFW provides an excellent new environment to build a CDSS that will prepare students for projects relevant to the NASA Human Research Program with support from contacts working on the Society for Design and Process Science (SDPS) [4]. This paper serves as inspiration and ground work for future grant preparation activity in the area of medical applications for cyber-physical systems.

Healthquest CDSS Project Contributing to NASA Digital Astronaut Project

The *Health Quest* CDSS project contributes to the NASA Digital Astronaut Project [3] by preparing university students and K-12 students for careers in Science Technology, Engineering, and Mathematics (STEM). The R&D workflow requirements promote learning in these areas. This is achieved by giving an opportunity for young K-12 students to observe how senior design and graduate students can build a working prototype of a CDSS by registering to an online platform managed by SDPS student chapter members.

Naturally the hands-on experience of architecting, designing, and developing the first system prototype at the university level strengthens and demonstrates their skills in STEM areas by motivating them to work on a project that extends an actual NASA project in the planning stages for the NASA Human Research Program, specifically the Digital Astronaut Project.

Healthquest CDSS Project Vision for STEM

The vision of the NASA project with STEM student work for Clinical Decision Support System (CDSS) is to engage both undergraduate and graduate students in the design/R&D of a medical guidance system for enhanced life support capability. This multi-university initiative, involving the Purdue System, UAB, Auburn, and other universities, can attract grants leveraging the medical resources available at UAB, ties to NASA Digital Astronaut Project and International Space Station (ISS), and the new Information Analytics and Visualization Center (IAV) at IPFW. Significant spillover benefits are expected to STEM initiatives via distance education and online/offline mentoring through the new IPFW and UAB student chapter organizations of the Society for Design and Process Science (SDPS) that can assist in the training of the students through e-learning. This approach is expected to generate a base of focused student activity that can supply enough qualified seniors and graduate students that can contribute to the modular development of the CDSS architecture. This paper reports on some collaborative work with UAB on the information architecture of a CDSS utilizing concept map technology for knowledge engineering of some of the modules using the Semantic Web. NASA outreach via STEM advanced by SDPS is expected to generate more qualified student researchers and freshman enrollment in computer science by stimulating interest and providing assistance both online and offline [5].

Healthquest CDSS Application

Astronauts experience many known and unknown challenges and biological stressors during space travel and in zero gravity that dramatically impact health. Even after the human body adapts to zero gravity, some obvious problems that have been observed to occur include disorientation, bone loss, cardiovascular deterioration, radiation exposure, and food poisoning/infections. In the case of (1) *Disorientation*: Zero gravity causes the body fluids to get disoriented and they tend to accumulate in the chest and the head, (2) *Cardiovascular deterioration*: bulging of the veins in the neck leading to loss of essential body nutrients, including a reduction in red blood cells. (3) *Bone loss*: The significant density reduction of the weight bearing bones causes osteoporosis, whose severity is accelerated due to prolonged stays in space [6]. (4) *Radiation*: Given the fact that the astronauts in space are no longer protected by the earth's magnetic field, the atoms bombard cells thereby penetrating and causing breaks in the DNA sequence which increases the risk for developing cancer. (5) *Food Poisoning/Infections*: Change in diet for the astronauts can lead to food poisoning and slight infections can create unknown deleterious consequences for the astronauts. This knowledge and appropriate countermeasures can be stored in a CDSS using concept maps.

Clinical Decision Support System as Solution for Automated Medical Support

Since it is impossible for a human medical expert to immediately respond with appropriate countermeasures to every medical problem in space, a CDSS could possibly improve the efficiency and response time to a particular emergency, in addition to providing longterm roadmaps to health that astronauts could follow. The CDSS should be able to provide the medical expert with automated recommendations via case based reasoning techniques combined with authenticated medical techniques available globally by accessing distributed knowledge bases, including knowledge stored as concept maps using Web Ontology Language (OWL) [7].

Importance of Concept Maps in Storing Information

Novak and Cañas [8] defined concept maps as "graphical tools for organizing and representing knowledge". These maps include concepts, usually enclosed in circles or boxes of some type, and relationships between concepts; such relationships are indicated by a line connecting two concepts [8]. A concept map allows a computer literate to illustrate an idea, concept, or domain knowledge to be described in a graphical form [9, 10]. Concept maps satisfy the conditions of meaningful learning because they provide a conceptually clear representation of comprehensive knowledge assimilated by a domain expert over time, which can be readily turned into XML and OWL instructions for machine processing [11]. The models generated with the use of concept maps can then be processed by the human mind as well as generate underlying specifications (e.g. in XML format) that can then be transformed into an executable format. Because the concept map provides an effictive methodology for determining user semantics, the loss of meaningful relationships involved in depicting domain information can potentially be minimized. Second, the concept map model can then be transformed to an executable format in an automated fashion using tools such as Cmaptools; this capability can further reduce the loss of semantics in process modeling.

Overview of Knowledge Engineering Approach Using Concept Maps

The proposed CDSS using concept maps for assisting medical experts should contain some of these basic components:

(i)   *Knowledge Base Module*. This component maintains the required domain information on space medicine in the form of concept maps which can later be converted into a suitable XML-based document.

(ii) *Knowledge Extraction Module*. This component is used to facilitate the extraction of relevant information from a database and the required knowledge from the knowledge base.

(iii) *Analysis Module*. This component analyzes the inputs provided by the Knowledge Extraction Module.

(iv) *User Interface Module*. This component handles the inputs on a particular health condition and provides the required information to medical experts.

The following steps are involved in the function of the CDSS:

(i) The medical experts build the knowledge base of diseases, abnormalities and their remedies based on previous experiences using concept maps. These concept maps are then converted into ontology documents.

(ii) The necessary information contained in the ontology document is extracted by the Knowledge Extraction module.

(iii) The medical experts provide the necessary inputs on a prevailing health abnormality of the astronauts in space using the user interface.

(iv) The Analysis Module accepts the information extracted by the Knowledge Extraction Module and clinical data provided by the flight surgeons using the User Interface Module.

(v) The medical experts either receive a particular remedy or a list of remedies for the health abnormality provided to the CDSS.

The knowledge base for the proposed CDSS is assembled based on past experiences of the medical experts and current medical research findings. The medical experts who have acquired years of experience working with the astronauts can readily impart their knowledge into a concept map using the Institute for Human–Machine Cognition (IHMC) CMaptools software. This tool simplifies the task of knowledge transfer in a way that can be machined processed via embedded concepts. The concept maps thus created will have the required information on diseases, their symptoms, and their required remedies.

The concept maps developed by CMaptools can also link the concepts to the relevant documents that can provide the medical expert with some additional information to derive a conclusion in proposing a remedy. The concept maps can be updated periodically to reflect the current knowledge on biological problems in space, in addition to remedies that have already been successfully tested. Since the concept maps are human perceivable from a display, a particular concept map can be readily reviewed and validated by a group of medical experts. Concept maps can be shared and linked together to form a single internetworked series of embedded concept maps to form a very large distributed knowledge base, similar to a neural network.

The knowledge base can have at least two primary components: (a) the visual concept map display and (b) the XML-based ontology file. While the concept map is in human perceivable format, the ontology file provides the machine processible form of information to the system. This information is extracted by the data and knowledge extraction module for identifying the necessary parts of information that can be used in deriving the final outcome and recommendations of the CDSS.

The Knowledge Extraction Module is primarily responsible for extracting the relevant information from the ontology document derived from the concept map. Usually the information extracted from this module is stored in a database which is used by the Analysis Module. This module uses an ontology parser to parse the ontology document under consideration.

The Analysis Module first gathers the required information extracted by the knowledge extraction module. It also gathers the clinical data of the astronauts in consideration. The module then performs an analysis of the clinical data with globally distributed knowledge to arrive at a list of conclusions that recommend solutions as a roadmap to health.

CDSS Grant Proposal: Visual Analytics

The primary medical research focus of the Information Analytics and Visualization (IAV) Center is conducted in the Healthcare Visual Analytics lab which can utilize the new Cave Automatic Virtual Environment (CAVE) technology to develop medical guidance systems using 3-D visual analytics. The CAVE can be used by trained STEM students to collect, analyze, and display data for clinicians, surgeons, and other domain experts when designing a functional CDSS. This type of medical guidance system can provide medical practitioners a technical roadmap with recommendations and rationale on best route of healthcare over a period of time, as well as emergency care. The various factors of input from both the patient and the world are synthesized to produce an output of recommendations displayed in a rich environment connected to remote knowledge sources and inference engines. The driver of this information processing is provided by the patient, detailing information on their health, and by the world, expediently reporting on advances in medical knowledge, building on the Unified Medical Language System (UMLS) [12] and recent advances in Web Services and Cloud technology.

The customized visual report and analysis of the data is provided by Artificial Intelligence (AI) modules with specific algorithms that can utilize available computation methods such as Matlab's Neural Networks and Fuzzy Logic toolbox. Semantic Web technology using concept maps and intelligent agents can be utilized to access global knowledge repositories to be translated from languages like Japanese and German to keep the system updated on most recent medical advances for automated processing and inferencing of new medical recommendations for users of the CDSS. Knowledge engineering techniques can be used to develop concept maps that store domain expertise for future recommendations and machine learning techniques utilizing Semantic Web and XML. Applications of this research range from automated drug discovery to onboard spaceflight medical diagnosis with collaboration opportunities existing with various companies, hospitals, and universities at the forefront of medical research.

The IAV Center will be able to utilize the advances made in the Digital Astronaut project, which is currently in the planning stages at NASA [3]. The comprehensive, mathematically based, digital representation of the astronaut will enable simulations to be conducted on space health under adverse conditions anticipated in space travel and under zero gravity. These detailed simulations of the astronaut's biological systems under various artificial stressors can expose unanticipated problems before they happen that can be addressed by a CDSS in advance.

The Digital Astronaut Project Extended by the Health Quest CDSS Project

The Digital Astronaut Project simulates human physiological function with enough transparency to formulate automated medical advice based on results from sensors that collect data and populate a rich database of dynamic health functions for an individual. This medical information can be analyzed automatically to produce a set of medical countermeasure recommendations for health optimization in microgravity gravity and conditions unlike Earth. However, the *Health Quest* CDSS system is an extension to the NASA project for conditions that occur in terrestrial gravity as well. Hence, STEM students have an opportunity to experience this type of project first-hand at the university level by developing a medical guidance system prototype that will prepare them for careers in medical informatics and software engineering.

SDPS STEM Online Portal Managing Project and K-12 Participation

Ongoing educational STEM activity can be managed by the online portal set up for the project similar to CPS virtual organization [13]. All IEEE documentation, design standards, implementation, and team collaborations will be done in this environment in full view of participating K-12 students so they can observe first-hand how senior design students can build a real working prototype of a CDSS system before graduation, according to prevailing software engineering industry standards. The online student portal can be managed by the STEM module of the SDPS student paper at IPFW. This organization serves as a platform to advance outreach activities via an interactive online portal, as a follow up to NASA outreach developed during the ASGC NASA Fellowship at Marshall Space Flight Center (2004–2006).

Health Quest Project Goals

Student led teams will be able to utilize industrial methodology for detailed design and implementation by developing the medical modules for the *Health Quest* system for full integration with other modules developed by several teams working in parallel according to a core architectural design based on IEEE standards.

STEM Experiential Design Process to Meet Industry Standards

Industry standards are important for senior design and graduate teams comprised of many student roles (e.g. project manager, chief architect, systems analysts, UML/SysML [1] designers, Object-oriented programmers). All phases of the IBM Rational Unified Process (RUP) [14] methodology with Axiomatic Design Process [15] using Software Engineering Body of Knowledge (SWEBOK) [16] and Project Management Body of Knowledge (PMBOK) [17] are studied, utilized, and executed from inception to transition phase, including formatted documentation according to IEEE 830 Software Requirements Specifications documentation, IEEE 1016 Software architecture/Design documentation, and IEEE 1058 Project Management Plan documentation. UML, SysML and the Unified Medical Language System (UMLS) [12] can be utilized for the detailed design phase. Tools include Institute for Human Machine Cognition (IHMC) Cmaptools [7] for project organization and version control, Visio/Telelogic [18] for application architecture and detailed design, Acclaro Design for Six Sigma (DFSS) for the information architecture, including hierarchical decomposition of functional requirements/ trade-off analysis/design matrix generating UML classes/dependency structure matrix (DSM) generating configurations of components and interfaces for the system architecture. Standard risk and project management techniques for validation and verification are developed, e.g. Quality Function Deployment (QFD), Failure Mode and Effects Analysis (FMEA), Probability Risk Assessment (PRA), Fault Tree Analysis (FTA), and Theory of Inventive Problem Solving (TRIZ) [19]. The Architecture Decomposition View (ADV) in Acclaro DFSS can be implemented with a Work Breakdown Structure (WBS) methodology in a standard tool like MS Project for Gantt charting and task assignments [20].

STEM Student Team Selection Process and Deliverables Format

The most qualified STEM students at IPFW may be identified and offered opportunities to receive funding to develop each phase managed through the portal and team collaboration tools (e.g. Basecamp). As deliverables, student teams may produce a project binder and CD for all work completed, by adapting the syllabus of the senior design course CS 360/460 sequence at IPFW. This comprehensive documentation approach will ensure that we can manage transition to the next team for continuation of the project in the SDPS student portal. The potential impact on total budget is provided by project phase according to RUP utilizing best practices for student teams (Table 4.1).

Table 4.1 Budget Impact by Phase of *Health Quest* CDSS Project with STEM Portal

Project phase	Narrative of deliverables reported to NASA via interactive SDPS/K-12portal
Initialization	Marketing includes showing team accomplishments and relationship of *Health Quest* system to the Digital Astronaut Project, Human Research Program, and ISS as the project develops in phases. Transparent online reporting of student roles assigned in each phase (Project manager, chief architect, designers, analysts, etc.) for K-12 understanding, including teaching software engineering concepts and tools to student teams involved, supervising development of *Health Quest* Project during RUP phases, managing development of K-12 student portal online, facilitating SDPS support for local student paper, managing IAV Center environment
Inception (RUP I)	CDSS Vision document, IHMC Concept Maps, Research, FMEA, Gantt Chart/progress reports, SWEBOK/PMBOK review, Power point presentations, internal website portal, team roles expansion from design to implementation
Elaboration (RUP II)	Application Architecture/Information Architecture/System Architecture, SRS-IEEE 830, SDD-1016, PMP-1058, QFD, PRA, FTA, TRIZ, Gantt Chart/progress reports, UML/SysML/UMLS, Power point presentations, external website portal K-12
Construction (RUP III)	CDSS *Health Quest* Project Code and internal/external portal finalization, team implementation role assignments (Programming, testing, integration, etc.) Gantt Chart/progress reports, Power point presentations
Transition (RUP IV)	Maintenance, SDPS documentation, Gantt Chart/progress reports, Power point presentations
Deployment	*Health Quest* Project finalization for CDSS and STEM portal prototype, project documentation, team results, STEM K-12 impact, and its relationship to the NASA Human Research Program to attract further support and resources from STEM, SDPS, IPFW, NASA and other stakeholders

NASA SMART Objectives Reporting

Every semester, reporting of results can be accomplished by concept maps on the Specific, Measureable, Attainable, Relevant, and Time-sensitive (SMART) goals aligned with NASA Objective 2.3 (Curricular Support Resources to Educate and Engage), focusing on NASA Education Outcome 1.1 and 1.2 (1st choice) [21]: (1) Specific goal of a teamwork initiative is achieved by attracting a group of talented university students for RUP and STEM implementation (2) Measurable goal of industrial standards is achieved by applying the detailed syllabus of the multi-semester senior design course (3) Attainable goal of intellectual empowerment is achieved using the SDPS STEM K-12 portal as a multi-disciplinary platform for idea cross-pollination and online mentorship (4) Relevant goal of increasing diverse computing enrollment is achieved since all types of STEM students can participate in module development (5) Time-sensitive goal of semester completion is achieved by meeting specific benchmarks specified by IEEE 1058 project management plan. These SMART objectives are achieved by building cross-disciplinary awareness among students, engaging STEM students in K-12 with experiential projects accessible online, involving SDPS for ongoing STEM resources, supporting student paper activity, inviting under-represented students to K-12 STEM portal/CDSS peripheral work, and teaching students software engineering concepts, principles, and tools according to prevailing industry standards.

Cyber-Physical Systems Design Environment

The Information Analytics and Visualization Center of Excellence at IPFW provides an excellent new facility for educational and training purposes in this emerging field of Cyber-physical systems applied to medical informatics, biomedical engineering, and healthcare engineering [13].

Health Quest CDSS as a CPS Entity

Cyber-physical systems involve the tight conjoining of software and systems elements with respect to dynamic component demands that require attention to timing concerns that can impact performance characteristics of concurrent, distributed, and synchronous systems. Students can be introduced to design methodologies such as Concurrent Object Modeling and Architectural Design Method (COMET) [22] for UML. CPS design concerns, such as sensors and actuators, can also be designed using SysML so that both hardware and software are specified at the same time. For example, the software developed for the *Health Quest* CDSS system needs to be designed with CPS principles for advanced embedded systems in mind. These types of systems require various hardware interfaces, e.g. sensors for data collection from the patient and actuators that respond automatically to the decisions made by the AI modules of the CDSS. This requires CPS analysis of the design process for the *Health Quest* CDSS, which is another activity that will be introduced in the STEM component of the project. The CDSS will then be able to make real-time decisions and recommendations based on artificial intelligence logic applied to the information flow from the patient.

SDPS Student Paper Collaboration

Student teams from other organization at universities interested in CPS development can participate. They can collaborate with the student teams at IPFW. This would require proper setup of social media and modular technology to support the distance education plan. Hence, a network of SDPS student papers can contribute modules to the *Health Quest* CDSS project core architecture as they form teams for collaboration. This approach provides a plan for national (and international) expansion as more student teams are added to the SDPS web portal online that promotes the Software Engineering Society (SES) [4]. As more student teams contribute modules to the *Health Quest* CDSS architecture utilizing concept maps and other methods, the project outcome becomes more robust. As a result, the system capability increases, impacting all three areas of Research, Outreach, and K-12 STEM education.

IPFW Information Analytics and Visualization Center

The new Information Analytics and Visualization Center (IAV) [23] can provide further support for student teams that wish to utilize advanced multimedia technology for CPS design visualization. The IAV Center was established in June 2010 with grant funds in the amount of $500,000 provided by *The Talent Initiative, Lilly Endowment Inc.* One of the eight labs is the Healthcare Visual Analytics Lab, which can be supported by other labs focusing on information visualization, intelligent systems, knowledge discovery and data mining, networking and security, scientific computation, semantic computing, and software engineering. The student teams have access to the following equipment, including a two-wall Cave Automatic Virtual Environments (CAVE) system composed of the following hardware:

- Cluster computer with 20 nodes, 160 cores
- SGI high computing graphics workstation
- Devices for wireless and sensor network
- Immersive 3D Vision Systems with tracking devices

Software Engineering Approach Guiding CDSS Development

Software engineering and project management are taught every year to senior design and graduate students utilizing projects as experiential case studies. In one project, the focus for the semester was demonstrating skills in software engineering and software project management with a case study (Project SESnet), which exercised skills in developing, integrating, documenting, and managing the modules for a website with various functionalities applying appropriate information technology. The course was designed to provide students with experience in task management essential and critical to the software industry. The same course syllabus can be used to educate and train students in software engineering so they can design a CDSS with cyberphysical specifications.

Table 4.2 NASA SMART goals showing *Health Quest* CDSS project input and output

Logic table: NASA goals	Inputs	Activities	Outputs	Outcomes	Outcome Measures
2.3 Curricular support resources (educate and engage) outcome 1.1/1.2 SMART goals Teamwork, Standards, intellectual empowerment, diverse enrollment, and semester completion	IPFW lab students K-12 students IAV lab INSGC funding SDPS student paper	R&D Medical Guidance System STEM outreach via SDPS student paper, mentoring online, Project Q&A module design/code proposals	Multiple (e.g. 4) teams of 5 IPFW students Each student mentors online 1–10 STEM students in K-12 Multiple CDSS designs and specifications/ coding for ISS	Outcome 1.1/1.2 Up to 200 students in STEM impacted/ semester More enrollment in computing and IAV R&D/NASA interest generated STEM e-learning support developed with virtual collaboration tools	Number of students impacted Quality of students measured by test score improvement to qualify for IAV Center/ NASA R&D work Surveys indicating increased interest in STEM careers Number of follow-up spinoff grants and papers

Table 4.3 Helpful websites used throughout CDSS development

PMI: http://marketplace.pmi.org Managing and Leading Software Projects: http://www.computer.org/portal/web/book_extras/fairley_software_projects and www.12207.com

Sommerville textbook: http://www.comp.lancs.ac.uk/computing/resources/IanS/SE8/index.html

UML and related papers: http://www.uml.org/

SWEBOK and IEEE industry standards: http://www.computer.org/portal/web/swebok

Concept map tool download: http://cmap.ihmc.us/download/

Visio UML/SysML stencil download: http://softwarestencils.com/uml/index.html

Axiomatic Design tool and related papers: http://www.axiomaticdesign.com/technology/papers.asp

Microsoft alliance (Visio, MS Project tools): http://msdn05.e-academy.com/purduefw_ece

Virtual collaboration tool: https://signup.37signals.com/basecamp/Free/signup/new

Quality printing and PDF conversion: http://www.cutepdf.com/

Joomla content management system (CMS): http://www.joomla.org/

IEEE standards and corporations: http://standards.ieee.org/develop/corpchan/mbrs1.html

Approach to Teach Software Engineering in Context of CPS

Appropriate textbooks and important references introduced students to fundamental concepts needed for understanding industry direction and practice (Table 4.2). Key reference links were provided in advance, in addition to periodic announcements in the Blackboard learning management system. By the end of the semester, each team was instructed to prepare for submission the following deliverables for the case study: an attractive team binder, CD e-documentation of all project and individual work in binder, and an online concept map with relevant attachments in RUP format that can be used to present to employers. Each student prepared an individual concept map also using Cmaptools that was linked to the main project Cmap containing all documentation for the project in progress. A detailed project description was provided to guide the student development according to established industry standards, e.g. IEEE 1058 for the project management plan (PMP), IEEE 830 for the software requirements specification (SRS), and IEEE 1016 for the software design description (SDD). The Project Management Body of Knowledge (PMBOK) and Software Engineering Body of Knowledge (SWEBOK), in addition to the systems engineering standards were applied as needed (Table 4.3).

RUP Phases of Development that can be Adapted to CDSS Design

For project SESnet, the RUP phases followed with SWEBOK standards can be adapted to software engineering of a CDSS in context of a cyber-physical system. During the RUP Inception Phase a formal vision document was prepared and various technologies were downloaded and setup for collaboration. Other documents and technologies were acquired and utilized in other phases throughout the semester, continuing with the Elaboration, Construction, and Transition phases. A concept map was used for organization, documentation, version control, presentations, and efficient file sharing.

The application architecture identified key modules for development and scoping which was referenced and updated throughout all the phases. The modular approach to web development started with an application architecture that organized the web platform into nine separate modules for development by the nine graduate students. Modules were individually tested after development and during integration. Scalability is achieved, since additional modules can be added at a later time with reconfigurable options, as more students volunteer to develop Project SESnet during the summer and beyond.

Throughout the project a Gantt chart using MS Project kept track of progress for assigned tasks based on the Project Management Plan (PMP) prepared according to standard IEEE 1058. Basecamp collaboration software was utilized for teamwork between classes, along with tools for expedient file sharing, such as www.dropbox.com. Joomla, Magento, and eFront and other content management system technologies were tested and integrated into the main Joomla site requiring authentication. Students participated in weekly meetings to present their integrated approach, while announcements of required documentation, assignments, and project work were posted online in Blackboard, which is the e-learning management system environment utilized at IPFW. Students developed a website and slide presentation for each module at the end of each RUP phase to report their progress, including describing any challenges encountered, problems solved, and plans made for the next phase.

Functional requirements of the various modules were assigned to each student who then utilized various tools, including, for example, Acclaro DFSS for dynamically capturing the information architecture using an axiomatic design approach, after the overall application architecture was established. A design matrix was generated with functional requirements that could be hierarchically decomposed for analysis, as needed. Nonfunctional requirements, such as performance needs for the system, were developed using Quality Function Deployment (QFD). Risk management was done by utilizing a design structure matrix (DSM) and Failure Mode and Effects Analysis (FMEA) techniques, which can be streamlined using the Acclaro tool for axiomatic design. Then UML and SysML are used for detailed design according to the analysis produced during the Elaboration Phase.

Following the Project Management Plan (PMP) as a guiding document (based on IEEE 1058), the project deadlines were established, assigned, and managed for various assigned tasks using a Gantt chart. The functional and nonfunctional requirements were archived in the software requirements specification (SRS) according to IEEE 830 standard by the systems engineer in the group managing the industrial documentation. Visio UML stencil download was utilized for the application architecture. The Application architecture, Information architecture, system architecture, and UML/SysML diagrams were archived according to the

IEEE 1016-2009 standard. Concept map technology provided by the Institute of Human Machine Cognition (IHMC) was utilized (Cmaptools) for version control and overall mapping of the project according to RUP phases. Tools such as these simplified record keeping (e.g., project work as well as homework) for the instructor, the team leader, and the individual members. This approach made it possible to educate students on PMBOK and SWEBOK theory using RUP and axiomatic design approach and support tools, while conforming to industry best practices for the production of a professional web platform.

During the Transition Phase, a duplicate staging site with identical modules was planned, allowing for the module testing without affecting progress of the main production site in development. In the IEEE 1016-2009 standard, each object oriented UML class structure was documented that depicted static and dynamic diagrams for modeling the case study based on the application and information architecture. Risk mitigation techniques were utilized (e.g. DSM, FMEA). Each student also managed his/her own Cmap with all assignments and project documentation provided as attachments according to the RUP phases.

Final Project Submission for CDSS

The final project submission includes (i) hardcopy binder of all project work completed for semester (ii) electronic copies of all work completed on CD (iii) electronic copies organized and displayed online using a concept map in HTML and CMap format, including all versions of documentation. Primary items in the Notebook Binder/CD/Cmap should contain: Table of contents with page numbers, overview of project, final version prints of documents (e.g. SRS/IEEE 830, PMP/IEEE 1058, SDD/IEEE 1016, Acclaro DFSS design matrix including FR/DP decomposition and other architectural views, Visio, Gantt, research, slides, optional software tests and coding), Cmap/slides (10 Slides/month, total 40-50 slides/Each team member can contribute to the total slide number.), exams, conclusion, references, and appendix (if needed). The following final items (including all versions) are required at minimum in the CD. The binder contains the final versions only. All the following items can be viewed from the team website:

1. Cover page and Table of Contents (including Concept map in HTML and Cmaptools format)
2. Brief Project Proposal and Research
3. Personal skills/interests sheets
4. Vision document
5. Architecture (all types in SDD document)

    a. Application architecture
    b. Information architecture
    c. System architecture
    d. UML architecture (structure diagrams with behavior diagrams)

6. SRS with top-level FR list from App Arch (IEEE 830 format)
7. Axiomatic design tool screenshots for any features used

    a. Design Matrix—FR/DP
    b. DSM—DP/DP
    c. FMEA
    d. QFD

8. Progress reports (in MS Word, table format, weekly from each member)
9. SWEBOK technical reviews KA-1 thru 11
10. PMP (IEEE 1058 format)
11. Gantt chart (comprehensive)
12. All slides
13. SDD (IEEE 1016 format)
14. Individual work (e.g. Exams I & II, HW, etc.)
15. UML (in the SDD document)
16. Appendix (Anything else stated in your PMP deliverables table, e.g. feasibility)
17. Key References, in any convenient format
18. Acknowledgements (optional)
19. Conclusion
20. Future work

Conclusion

We proposed a university student team led *Health Quest* project that develops a clinical decision support system (CDSS) with interactive STEM online portal (K-12) as an extension to the NASA Digital Astronaut Project in context of a Cyber-physical system. This medical guidance system will be built by both graduate and undergraduate students, who can provide guidance/mentorship to K-12 students online throughout all development phases. The Information Analytics and Visualization (IAV) Center at IPFW provides an excellent new environment to build a CDSS that will prepare students for projects relevant to the NASA Human Research Program with support from contacts working on the Society for Design and Process Science (SDPS) [4].

Acknowledgments I would like to thank Dr. Albayyari (IPFW Associate Vice Chancellor), Dr. Ng (IPFW Chair of the CS Department and primary mentor), Dr. Moradi (UAB Director of the Center for Biophysical Sciences and Engineering), Dr. Nowak (NISTEM director at IPFW), and Dr. Marghitu (SDPS co-chair of Student paper Development), for providing reference letters to NASA in support of the *Health Quest* CDSS Project. I also would like to thank Dr. Abdullah Eroglu, Dr. John Carbone, Dr. Varadraj Prabhu, Dr. Sang Suh, Dr. Murat M. Tanik, Dr. Hiroshi Yamaguchi, Dr. Bernd Kramer, Dr. Scott Brande, and Dr. Stan Gatchel for their advice, guidance, and encouragement of our SDPS/SES efforts, especially in promoting STEM and CPS development via SDPS student papers.

References

1. UML/SysMLhttp://www.uml.org/ [accessed March 16, 2012]
2. Science Engineering Technology Math coalition: http://www.stemedcoalition.org/ [Accessed October 3, 2012]
3. NASA Digital Astronaut Project, http://spaceflightsystems.grc.nasa.gov/SOPO/ICHO/HRP/DA/ [accessed March 16, 2012]
4. SDPS/SES www.sdpsnet.org [accessed March 16, 2012]
5. Gurupur V., Moradi L. G., and Tanik U. J. (2011). Information Architecture of a Clinical Decision Support System for NASA Life Support Project Advancing STEM. SDPS—2011.
6. NASA related education [Online]. Available: http://www.astrophys-assist.com/educate/spaceflight/spaceflight.htm. [Accessed April 14, 2011]
7. Human Machine Cognition (IHMC) Cmaptools http://cmap.ihmc.us/download/ [accessed March 16, 2012]
8. Novak J.D. and Cañas A. J. (2006). The theory underlying Concept Maps and how to construct and use them [Online]. Available: http://cmap.ihmc.us/Publications/ResearchPapers/TheoryUnderlyingConceptMaps.pdf. [Accessed: April 7, 2011]
9. Cañas A. J., Novak J.D., and F. M. González (2004). Two-layered approach to knowledge representation using conceptual maps and description logics [Online]. Available: http://cmc.ihmc.us/papers/cmc2004-205.pdf. [Accessed: April 8, 2011]
10. Gurupur V. and Tanik M.M., "A System for Building Clinical Research Applications using Semantic Web-Based Approach," Journal of Medical Systems, DOI: 10.1007/s10916-010-9445-8.
11. Gurupur V., 2010, "A Framework for Composite Service Development: Process-as-a-Concept," Ph.D. dissertation, Department of Electrical and Computer Engineering; University of alabama at Birmingham, Birmingham, AL.
12. Unified Medical Language System http://www.nlm.nih.gov/research/umls/ [accessed March 16, 2012]
13. CPS http://cps-vo.org/ [accessed March 16, 2012]
14. IBM Rational Unified Process (RUP) http://www.sei.cmu.edu/risk/ [accessed March 16, 2012]
15. Axiomatic Design Process http://www.axiomaticdesign.com/technology/papers.asp [accessed March 16, 2012]
16. Software Engineering Body of Knowledge (SWEBOK) http://www.computer.org/portal/web/swebok [accessed March 16, 2012]
17. Project Management Body of Knowledge (PMBOK) http://marketplace.pmi.org [accessed March 16, 2012]
18. Visio/Telelogic (UML tools) [accessed March 16, 2012] http://msdn05.e-academy.com/purduefw_ece
19. Risk management http://www.sei.cmu.edu/risk/ [accessed March 16, 2012]
20. Managing and Leading Software Projects by Richard Fairley, Wiley,2009 (ISBN: 978-0-470-40573-4): http://www.computer.org/portal/web/book_extras/fairley_software_projects and www.12207.com [accessed March 16, 2012]
21. NASA grants https://engineering.purdue.edu/INSGC [accessed March 16, 2012]
22. COMET [accessed March 16, 2012]
23. IAV Center http://www.insideindianabusiness.com/newsitem.asp?ID=41933 [accessed March 16, 2012]

Chapter 5
Radically Simplifying Cyber Security

Dan Kruger and John N. Carbone

Introduction

Cyber security professionals are publicly acknowledging [1, 2], that the traditional approaches to securing information fail because the threat environment has become impossibly complex. This chapter on a fundamentally new approach to securing information that can radically simplify cyber security.

The definition of the problem is the problem. Cyber security has become unmanageably complex because the definitions of security do not match the operational environment—and they haven't for a long time. Historical approaches to securing information were more applicable in the mainframe era where systems were more self contained, independent, and didn't touch many external systems as they do in the web era today. Hence, the following security definitions were in use then are still in force today:

Perimeters: Information is protected when you control access to it by establishing a perimeter. Defining the perimeter was simple—control physical access to the building and the terminals. As an example, years ago an information systems director of a major oil company became nearly apoplectic when personal computers were introduced to his company. He said, "Once you let them in, we'll never have control of our data again." He was right.

Processing Capacity: For information to be readily useful it needs to be unprotected—stored "in the clear." Protecting information takes significant processor capacity and makes it more time consuming and difficult to process.

Training: Keep users from leaking information by training them to understand and follow the organization's security policies.

D. Kruger (✉)
Absio Corporation, 8321 S. Sangre De Cristo Rd, Ste 302, Littleton CO 80127-6426, USA
e-mail: dan.kruger@absio.com

J. N. Carbone
2730 Woods Lane, Garland TX 75044, USA
e-mail: jcarbone@raytheon.com

S. C. Suh et al. (eds.), *Applied Cyber-Physical Systems*,
DOI: 10.1007/978-1-4614-7336-7_5,
© Springer Science+Business Media New York 2014

Therefore, this paper addresses the challenge of fundamentally enhancing and changing existing security paradigms by minimizing system ambiguities and significantly simplifying the cyber security of information content.

Perimeter Ambiguity

Historical assumptions regarding perimeters are no longer valid today; starting with the notion that one can simply define a perimeter around the complex system interconnects that currently exist. It's worthwhile to look at just how hard it is to merely define the perimeter—much less defend it. Begin with a mainframe and terminals in a building you control. Then add PCs, each with their own processors, local storage, I/O ports and local applications; each PC has a perimeter of its own. Then add portable computers designed to go outside the physical perimeter. Subsequently adding a local area network, adds physical perimeter elements you must now defend which can be prone to tapping and possible eavesdropping. Lastly, adding wide area networks now expands your perimeter elements to include somebody else's wires and switches.

These definitions describe distributed computing in the mid 1980s—the last time anybody was able to do a marginally credible job of defining a perimeter— and this is where the notions of cyber security have been mired in seemingly outdated principles for today's complex system environments. It is well known in the cyber security field to discuss protecting networks as if that were synonymous with protecting information. Network protection, even if perfect, is a partial security solution at best.

Next we make the move to the mid 1990s and add the commercialized Internet. Then add ubiquitous email and the World Wide Web, both enabling the unconstrained distribution of information by unauthenticated users. Next add high-density portable drives and make some of them smaller than a postage stamp. Then add radios in multiple flavors—cellular, Wi-Fi and Bluetooth, with more on the way. Include applications, too—peer-to-peer networking, streaming media and social networking. Can any organization actually define its perimeter? If it were possible, would you be able to defend it? Let's look at the complexity of defending a small workgroup.

Every digital device, operating system (OS), application, transmission path, file and human being has multiple attack vectors (paths to the target) and attack surfaces (that which can be attacked). When devices are booted, users log in, launch applications, open emails, browse websites, manipulate files or transmit information, those actions open vectors and expose surfaces, creating attack opportunities.

Let's do some rough math: 10 users $\times$ 3 devices each $\times$ 20 applications each $\times$ 5 attack vectors $\times$ 100 interactions daily. That's 3,000 perimeter elements to defend and 300,000 threat opportunities a day (Argue the math if you want—it's actually worse than the example.). If that perimeter could be precisely defined, who would have the time, money and expertise to close all of the holes and keep them closed?

In addition to sheer numbers, threats are unpredictable and dynamically complex. Addressing one can produce unintended consequences in others, and nifty new technologies compound the problem. It's only a matter of time before we hear something like this: "A major breach was traced back to an exploit that used the smart refrigerator interface in the CIO's home network to infect his tablet, which then invaded his company's network."

Even if you could define all of the perimeters, they are too complex to defend. This is not a call to abandon the perimeter; the better the perimeter defense, the more sophisticated the attacker needs to be. But it is clear that the current definition of the perimeter is only the outer layer of defense [3–5]. Therefore, defense in depth is required.

Processing Capacity

If information protection is properly engineered, there is more than enough processor capacity to protect information everywhere it's at rest or in motion—if you include the processor on edge devices. It's critical to include edge device capacity in any approach to cyber security because increasingly sophisticated edge devices:

- Are where most information lives
- Have I/O ports for exporting information
- Are often portable and easily stolen or captured
- Are what users are using and will continue to use

That brings us to the cloud. The drive toward clouds and the drive for more powerful mobile devices are, in many ways, at odds with each other. Cloud computing assumes that processing is done in the cloud and that edge devices host what are essentially visually pleasing dumb terminal emulations that always have a good connection. We have recreated the mainframe model in an attempt to define the perimeter. The conceptual security advantages of clouds are obvious: They have a definable perimeter. But the perimeter itself raises a fundamental security problem [6].

If users can take the product of cloud computing and freely distribute it, then cloud security stops at the perimeter of the cloud (more precisely, at the data-center's interface to the Internet). Information security solutions must comprehend information security from cloud to cloud, cloud to edge device, and edge device to edge device. Information security must persist.

Disable-your-hardware solutions for securing cloud connections have not and are not going to be fully implemented. People are not going to disable their devices' I/O ports. Users will store and share information on their edge devices regardless of security policies.

People and Training

The third assumption was always more hope than reality. Do we expect users to follow security policies that make their jobs more difficult than they already are— especially when they know the policies make little difference? Do we expect that good training will stop all social engineering and/or thwart suborned or malicious insiders? Stop outsiders who masquerade as insiders? Do we expect that sufficient training will ensure that every user in the distribution chain will make the correct security decisions about a piece of information [7]? If the perimeter has been breached and the network has been invaded. We cannot stop it. What can we do? We must, "Change the game".

Methods

Enhanced Security Via Separation of Knowledge and Context

Knowledge and contextual understanding of it has been debated for years. Brillouin [8] defined knowledge succinctly as resulting from a certain amount of thinking and distinct from information which had no value, was the "result of choice", and was the raw material consisting of a mere collection of data. Additionally, Brillouin concluded that a hundred random sentences from a newspaper, or a line of Shakespeare, or even a theorem of Einstein have exactly the same information value. Therefore, information content has "no value" until it has been thought about and thus turned into knowledge. Following this train of thought, knowledge is created through the amount of context, which can be recombinantly assimilated over time until a threshold of relationship understanding is achieved [9]. Gruber [10] states that collective intelligence emerges if data collected from all people is aggregated and "recombined" to create new knowledge. To form an understanding of the relationship between different knowledge and contexts when assimilating knowledge, the associated relationships can be written symbolically as knowledge Ki and the associated context relationship Rj where, $Ki(Rj)$ represents a recombination of knowledge and context as shown in Equation 1 below.

For preventing cyber attackers this is a key understanding since the amount of context received is a function of how much access can be achieved. The amount of access obviously increases the context and the possibility of damage. The more an attacker knows the more he can do. Thus, to remain secure we need to separate the knowledge and information content and hence negate an attackers ability to gain context. Hence, the next sections discuss the simplified mechanisms to enhance cyber security by obfuscating the knowledge and context itself.

$$\sum_{i,j} K_i(\ R_j)$$

Equation 1, Recombinant Knowledge Assimilation

Information that Protects Itself

Cyber security is radically simplified if we move primary information security into the information itself. With today's processors, storage density and the right engineering, it is possible to make every piece of information a "hard target" that protects itself and is still easy to use. The focus of cyber security can shift from the utterly impossible (defending the undefendable perimeter and persuading the unpersuadable user) to the merely difficult (moving information out of the clear). Hence, if information can protect itself, the assumptions behind cyber security are very different (Table 5.1).

Establishing Persistent Distribution Control
at the Object Level

The mechanism to obfuscate knowledge by securing contextual information content is performed by the creation of an individually protected object called a SEGCO—a Secure Extensible Global Content Object. As shown in Fig. 5.1, the application architecture supporting the creation, storage, transportation and authentication of SEGCOs, along with the formalized application programming interface (API), thus making it available to developers, is called the Persistent Distribution Control System (PDCS). PDCS enables developers to build persistent distribution control into their applications, which enables the encapsulation of information within SEGCOs as any new information is created, enables storage and transmission of only SEGCOs and hence, creates a secure environment where information at rest and in motion is dynamically secured. If an attacker actually

Table 5.1 Changing cyber security assumptions

Current assumptions	New assumptions
Information is protected when you control access to it by establishing a perimeter	Information is protected when it is bound to persistent distribution controls and is never in the clear except when actually being used
Information needs to be stored in the clear to be easy to use	Information does not need to be stored in the clear to be easy to use
You keep users from leaking information by establishing rules and hoping they follow them	You keep users from leaking information by establishing rules that their applications enforce

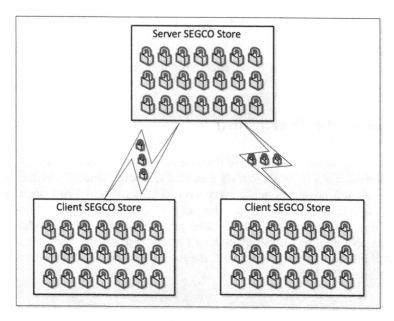

Fig. 5.1 Secure extensible global content objects

succeeds in breaching a client device, server or signal, the only content they achieve access to is a mountain of individually secured SEGCOs.

Key SEGCO/PDCS attributes include the following:

- SEGCOs are individually encrypted, each with a distinct key. That radically reduces the number of useful attack vectors and offers the most resistant of all attack surfaces. Attacks designed to copy and illicitly export data fail to deliver information attackers can use.
- PDCS stores SEGCOs uniformly. SEGCOs are indistinguishable from each other. This makes getting the right information object analogous to finding a particular grain of sand on a beach of identical grains of sand. The more grains of sand, the harder the problem for attackers.
- SEGCOs are transmitted in an encrypted tunnel. A breached tunnel yields only SEGCO's.
- PDCS and SEGCOs make it possible to build tools and applications that work across almost any kind of hardware and OS. Information that is secured in a SEGCO from the moment it is created will be safe in motion and at rest wherever it exists.
- SEGCOs and PDCS make it possible to build mechanisms to authenticate users, devices and applications prior to decrypting the information (chain-of-trust fingerprinting).

Therefore, the consistency of SEGCO metadata and the audit function of PDCS make it possible to implement auditing systems that monitor the movement and use of information in near real time. Those systems may be able to flag the misuse of information objects fast enough to keep the attack from succeeding and the comprehensiveness and immutability of the audit trail can establish a legally admissible chain of custody to aid in prosecution.

Finally, SEGCOs and PDCS can therefore provide an array of capabilities that application developers could use to create new solutions, including:

- Fine-grained control of secondary distribution such as forward, export, copy/ paste, or print.
- Secure documents that redact themselves as they move through distribution.
- Cross-domain secure collaboration without inter-domain access.
- Copyright protection that does not unduly restrict the user's access to content.
- Commercial authentication without risk of exposing personally identifiable information.

SEGCO-PDCS Requirements and Architecture

Fundamentally, a SEGCO must: be platform and content-agnostic (support any data type) as shown in Fig. 5.2, be distinctly encrypted (provide a different key), remain encrypted in motion and at rest, contain distribution control information inside the encrypted envelope, contain audit information inside the encrypted envelope, not indicate the type of content it contains and be randomly and/or nonsensically named.

Subsequently, PDCS must be implemented using client–server architecture to disperse the encryption/decryption processor load across millions of edge processors and to provide a common multiplatform security API as shown in Fig. 5.3. This point is critical since developers rarely have the knowledge required to build information security into their applications [11, 12], and it would likely do more

Fig. 5.2 SEGCO-PDCS
architecture

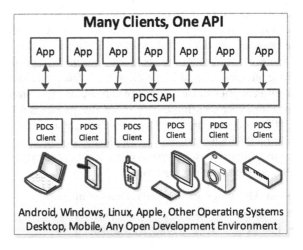

Fig. 5.3 PDCS architecture

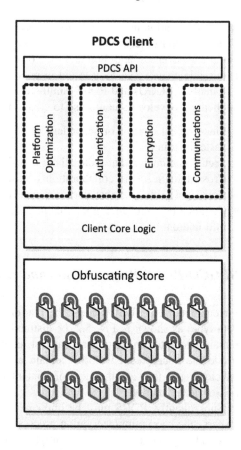

harm than good if their security solutions were all different. Methods and developer tools for securing information objects must be easy to use, standardized, broadly applicable and inexpensive. They should also support disconnected use, enable secure storage on edge devices and servers, and enable hierarchical server architecture in order to place servers into the resource-loaded environment where they are needed. From a communications or data-in-motion perspective the architecture should enable secure transport and interim storage for applications that require large amounts of data in the clear (e.g. cloud to cloud, cloud to edge) and enable intelligent prioritization of bandwidth not just for communications optimization but also for cyber security. Empty space always leaves room for infiltration.

PDCS clients must be OS-optimized, authentication method agnostic, encryption method agnostic and communications method agnostic (support any Internet protocol transmission path) so that clients can run efficiently on any platform. The PDCS client architecture provides:

- An API with common calls across platforms.
- Platform-specific performance optimization.
- Platform-specific module for authentication.

- Platform-specific module for encryption.
- Platform-specific module for communications.
- Common core logic across platforms.
- A SEGCO store that makes SEGCOs uniform.

PDCS content servers should be as naïve and lightweight as possible—and they can be since client devices are performing application processing and "heavy lifting" such as encrypt/decrypt. The key functions of PDCS servers will be to route, store and forward SEGCO's, while limiting access to authenticated users and devices (no support for system level anonymity) and to provide a mesh network set of management functions because of their distributed nature and finally, to provide the necessary resilience and redundancy components necessary for seamless disaster recovery functionality.

Results

Outcomes of SEGCOs and Persistent Distribution Control

What current hard problems would be simplified if applications stored and moved information in SEGCOs managed through a PDCS application architecture? These are some we have identified:

- Insider threat: If Bradley copies 200,000 SEGCOs to a Lady Gaga CD and gives them to Julian, Julian will have a useless CD—and nothing to publish.
- Malware-based storage attacks: If malware is able to convey only SEGCOs to outsiders, the malware is of no value to the malware writer or purveyor.
- Signal intercepts (man-in-the-middle attack): If a signal is intercepted and the signal contains only a stream of SEGCOs, the intercept is of no value to the eavesdropper.
- Device capture: If a device containing thousands or more SEGCOs is captured, the device has no value to the thief.
- Personal identity protection: Applications can be built that enable the complete separation of personally identifiable information from the information objects that represent the person.
- Cross-domain information sharing without cross-domain access: Applications can be built that support software-based secure intermediate logical networks.
- Copyright preservation without undue restrictions on usability: Applications can be built that ensure copyright holders can track and get paid for their content while allowing buyers to access their content on whatever device they are using at the moment.
- Intelligent traffic prioritization: Applications can be built that support client-based prioritization of traffic—a critical need anywhere bandwidth is scarce and urgency is high.

Remaining Vulnerabilities and Next Steps

It should be noted that persistent distribution control does not eliminate information security problems. However, PDCS shrinks the problem considerably by undermining the value of storage and eavesdropping exploits, reducing the requirement for users to follow security policies and increasing the risk, time, and effort of attempting exploits.

Hence, full information security requires a comprehensive combination of tactics for closing the holes in our security of our hardware and software supply chains and "at the bottom of the stack" [13]. A number of solutions are needed, such as separation kernels, secure operating systems and associated access to them, encrypted and dynamically allocated memory, effective micro virtual machines with discrete levels of focused capability, robust chain-of-trust solutions based upon pedigree creation, modification and deletion. Last but not least a comprehensive roadmap is needed for achieving a full spectrum of capabilities security nexus and can be summarized in these suggested next steps: reduce the attack surface, validate trust and verify your user base, continuously evaluate your enemy's methodologies, dynamically discover and vanquish the intruders, measure, automate, and audit comprehensively, vary your methods, processes, and locations to make yourself a moving target, and most importantly grow the vitality of your force through proper preparation, education and focused training.

Conclusions

This chapter addressed the challenge of fundamentally enhancing and changing existing security paradigms by minimizing system ambiguities and significantly simplifying the cyber security of information content. Thus, by shrinking the size and scope of the threat universe, which cyber professionals describe as reducing the attack surface, the PDCS described herein showed how cyber security professionals and the industry can focus on the small number of truly sophisticated attacks and attackers while simultaneously reducing the time needed to address nuisance attacks. The SEGCO mechanisms and PDCS architecture described showed how cyber attackers will be forced to have more expertise, expend more capital and take on more risk since attacking systems secured with these formalized methods will become much more expensive and have a much smaller chance of success. Therefore, information that can protect itself ruins the economics of hacking.

References

1. D. Barrett, "U.S. Outgunned in Hacker War", Wall Street Journal, March 28, 2012.
2. ThreatPost, Kapersky Labs: Experts Tell Senate: Government Networks Owned, Resistance Is Futile, March 21, 2012
3. Axel Buecker, Per Andreas and Scott Paisley, "Understanding IT Perimeter Security," IBM Corporation, 2008.
4. Marcia Savage, "Perimeter Defenses Deemed Ineffective Against Modern Security Threats," Information Security, June 30, 2010.
5. Simson Garfinkel, "The Deperimeter Problem," CSO Online, November 1, 2005.
6. Anh Nguyen, "Infosec: Cloud Computing 'Explodes' the Security Perimeter," CSO Online, April 25, 2011.
7. "Top Cause of Data Breaches? Negligent Insiders," Help Net Security, Ponemon Institute, March 22, 2012.
8. L. Brillouin, Science and information theory: Dover, 2004.
9. Crowder, J. A., Carbone, J. N., "The Great Migration: Information to Knowledge using Cognition-Based Frameworks." Springer Science, New York (2011).
10. T. Gruber, "Collective knowledge systems: Where the social web meets the semantic web," Web Semantics: Science, Services and Agents on the World Wide Web, vol. 6, pp. 4–13, 2008.
11. Adam Cummings and Ron Bendes "Information Security in Application Development Projects," cmu95752, Carnegie Mellon University, March 7, 2012
12. "Risk Across the Phases of Application Security," Help Net Security, Ponemon Institute, March 21, 2012.
13. A. Metke, "Security technology for smart grid networks" IEEE Transactions on Smart Grid, 2010.

Chapter 6
Cyber-Physical System Architectures for Dynamic, Real-Time "Need-to-Know" Authorization

James A. Crowder and John N. Carbone

Introduction

Information systems for securing confidential/classified information have typically relied on a manually compiled Access Control List (ACL) to control access to information. Each item of classified/confidential information is associated with an ACL. This insures a level of security and can be accessed by anyone who has been authorized. Attempts to overcome problems associated with ACLs are Role-Based Access Control (RBAC) and Attribute-Based Access Control (ABAC) [12]. RBAC has crucial security weaknesses in that an authorized user may be able to look at the entire container, whereas he only has need for a particular part of the information, based on context. Although, the data are at the same classification level, a user may not need to know (or should not know) things like financial or location information. This is called "Contextual-Need-to-Know" (CNK). Even in ABAC, where access is granted not based on the rights of the subject associated with a user after authentication, but based on attributes of the user, the user has to prove claims about his attributes to the access control engine. Again, this method is problematic in a real-time, dynamically changing environment like a battlefield or war-time situation, where access needs may change as the situational parameters change.

In short, current access control methodologies are inflexible and do not allow for frequent, real-time changes to classification levels, needs-to-know designations, based on combinations of information and contextual need-to-know indictors. Here, a multi-agent CNK methodology is proposed that takes advantage of classification and quantum fractal encryption algorithms for real-time authorization requests to classified information. Instead of course-grained ACLs, a multiple, intelligent

J. A. Crowder (✉) · J. N. Carbone
Raytheon Intelligence, Information, and Systems, Garland, TX 75042, USA
e-mail: JACrowder@raytheon.com

J. N. Carbone
e-mail: jcarbone@raytheon.com

S. C. Suh et al. (eds.), *Applied Cyber-Physical Systems*,
DOI: 10.1007/978-1-4614-7336-7_6,
© Springer Science+Business Media New York 2014

information agent system is proposed that will analyze the context of the user, the information, and context of the information (including the context of combined information that may change the security level), along with the current situational parameters surrounding the request and informational context. These are combined with the user's clearances and need-to-know criteria to determine what information and context may be shared with the user. This new system makes it possible to define, assign, and enforce security policies that may change real-time as the intelligent information agents will learn and adapt the information and CNK, based on contextual needs and availabilities. This is combined with a multi-dimensional quantum fractal encryption scheme which can "unpack" information at any security, context, information, and need-to-know combination.

New Approaches to Access Control

a. Typical Need-to-Know Scenario

The underlying assumption for current systems is that all classified information is stored in a server (or set of servers) as illustrated in Fig. 6.1. The classified information requested is labeled "C" in Fig. 6.1 [8]. The information needed may not be the entire contents of "C." In order to access the information, the user must get authorization from the access control system (e.g., LDAP). With authorization,

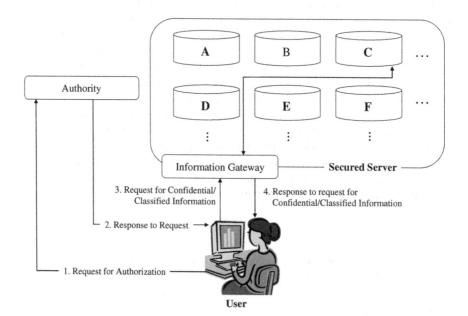

Fig. 6.1 Typical need-to-know authorization

based on some access control methodology (e.g., RBAC), the user can now access the information. However, even with tight row and element-based security tagging (e.g., Oracle data tagging), need-to-know is hard to manage since real-time changes required, as situational awareness requirements and conditions change, require manual changes to access control lists and individual data tagging [13].

Some of the issues associated with current access control methodologies (and their solutions with our new architecture) involve:

Distributed Information Sources

Information sources available on-line are inherently distributed. Furthermore, these sources typically are of different modalities. This may mean maintaining multiple access lists. This can be overcome with an architecture consisting of multiple, intelligent information agents specialized for different heterogeneous information sources.

Data Sharability

User requests may need to access services, resources, and information in an asynchronous manner in order to support a variety of tasks and requests, it is difficult for current systems to assess the viability of the "need-to-know" aspect of the response to the users request in a real-time, asynchronous manner as the information is being assembled across a variety of resources and services. It is desirable to have a multi-agent system that communicates between agents inside an architecture that supports sharability of Intelligent Information Agent (I^2A) capabilities and retrieves information in order to assess CNK before any information in presented to the user [1].

Flexibility

Current systems cannot handle on-demand changes to need-to-know capabilities and multiple-security CKN based on asynchronous information retrieval scenarios [9]. A multiple I^2A system can interact with new configuration and informational awareness scenarios and information to provide "on-demand" changes that support changing information requirements for decision making in real-time warfighter situations.

Robustness

Current systems that are manually manipulated and maintained are inherently brittle [11]. When information and control is distributed with an I^2A architecture, the system is able to degrade gracefully, even when some of the agents are out of service temporarily. This has a real advantage in a dynamic system [2].

Quality of Authorization Information

The possible existence of overlapping authorization information items from multiple sources can pose problems for hard-coded systems [10]. However the use of I^2A for information sharing and cross-validation enhances correctness of data. The I^2A architecture allows agents to interact, negotiate, and find the most accurate data.

b. CKN Utilizing the I^2A Framework

Figure 6.2 illustrates an initial look at a multi-intelligent information agent framework for real-time need-to-know scenarios [8].

In order to accommodate real-time changes in situational conditions, the Intelligent Information Agents utilize context sensitive, Fuzzy, Self-Organizing Topical Maps with the following properties:

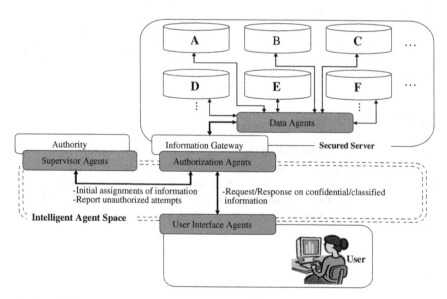

Fig. 6.2 Notional architecture for CKN utilizing the I^2A framework

Coverage. The Intelligent Information Agents must be able to protect all classified information according to given security policies, in addition to additional contextual need-to-know criteria [4].

Adaptiveness. The Intelligent Information Agents must be flexible enough to adapt to frequent changes in situational awareness to hand dynamic CKN for various types of information:

- Text
- Graphics
- Audio
- Video
- Etc.

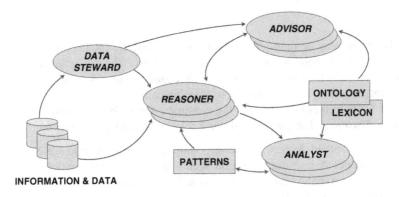

Fig. 6.3 CKN multi-agent framework

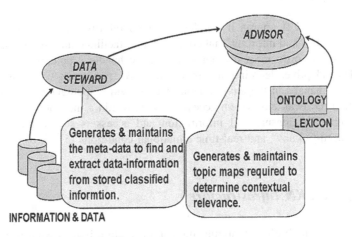

Fig. 6.4 The data Steward and advisor agents

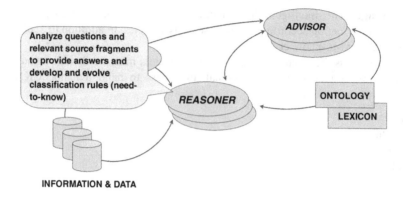

Fig. 6.5 The reasoner agent

Also, there may be multiple levels of changes and the Intelligent Information Agents must be able to adapt. Figure 6.3 shows the types of Intelligent Information Agents utilized within the architecture. Figures 6.4 and 6.5 describe the agents and their roles within the I^2A architecture.

This CNK framework provides the constructs for a *Context-Sensitive Authorization Reasoner*. The purposes of these intelligent software agents are [3]:

- The Data Steward Agents are the informational interface for the CKN system.
- The Reasoner Agents (authorization agents) make the determination of CKN.
- The Analyst Agents compute the contextual closeness of the requestor's role and request to the information content.
- The Advisor Agents make the final recommendations as to the approved informational access.

The multiple, I^2A approach handles classified information utilizing multiple threads of control and informational cooperation to facilitate the use of distributed resources and informational awareness criteria to handle a potentially changing CKN. The adaptive, learning nature of the I^2A framework can manage the response to changes in the environment. This provides spontaneous mapping of information retrieval and system access requests within a multiple security level environment. This ensures all information and access is secure regardless of security, need-to-know, and real-time situational conditions [5].

c. Adaptive CKN Infrastructure

The underlying multiple I^2A architectural infrastructure is illustrated in Fig. 6.6. This allows the intelligent agents to learn and adapt to changing environments, and facilitates cooperation and information sharing between intelligent agents [5].

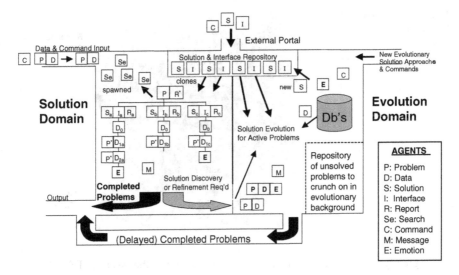

Fig. 6.6 The I2A processing infrastructure

The goal here is to provide a highly-reliable authorization methodology and architecture for multiple levels of classification within a real-time, dynamic security environment that can support battlefield situations. The agents form solution chains for each informational retrieval that corresponds to a given request. Each leg of the solution may have different classifications, based on information content, context, and need-to-know. An overall evaluation of the total information CKN is made and determination as to which (if any) of the information can be made available to the requestor. Below is a discussion of the multi-dimensional encrypting methodology utilized to unpack (provide) the information that is allowed to the user. The underlying technology that makes the analysis possible is the Context-Sensitive, Fuzzy, Self-Organizing Topical Map, along with the Analyst and Reasoner Information Agents.

d. The Fuzzy, Self-Organizing Topical Map

The topical maps for the $\mathbf{I^2A}$ architecture are loosely, or "fuzzily," correlated sets of information that define topical relations between these data. The starting point for building a Topic Map is the Fuzzy, Self-Organizing Map (FSOM). The FSOM is a general method for analyzing and visualizing complex, multidimensional data. It has been used to order information into an informational 'meaning' map such that similar information lies next to one another in a two dimensional grid [6]–03.

The CKM FSOM is actually built from two separate FSOMs. The first is a semantic FSOM that organizes the information grammatically and semantically into categories used to encode the inputted information as a histogram [7].

Fig. 6.7 Depiction of the fuzzy, self-organizing topical map

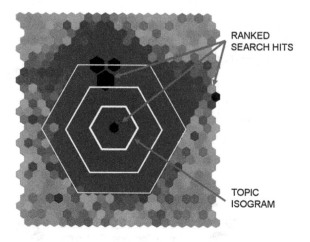

This greatly reduces the dimensionality, simplifying the second FSOM, the Information Map. The Information Map is labeled to form a Topic Map that has several important attributes:

- Intelligent agents can explore the map for information located by meaning.
- Searches use contextual information to find links to relevant information within other source documents.
- The Information Map is self-maintaining and automatically locates input on the grid, 'unsupervised'.

  Figure 6.7 illustrates an FSOM with information search hits superimposed. The larger hexagons denote information sources that best fit the search criterion. The isograms denote how close the hits are to particular information topics. Once the FSOM has been developed, it can be enhanced to include a Topic Map, shown in Fig. 6.8.

  In Fig. 6.7, the larger hexagons denote informational sources that best fit the criterion. The isograms denote how close the hits are to a particular target, and thus the contextual relevancy to the requestor's role to the topical information requested. This approach to mapping information by "contextual meaning" avoids problems with classical language or rule-based methods. The multi-agent association framework is illustrated in Fig. 6.9.

Multi-Dimensional Quantum Fractal Encryption

In order to facilitate providing CKN across multiple security levels within the same system or framework, and encryption scheme that allow information to be encrypted at any combination of information, context, content, and need-to-know is required. The Multi-dimensional fractal matrix defines real-time a multi-dimensional fractal

Fig. 6.8 The CKN topical map structure

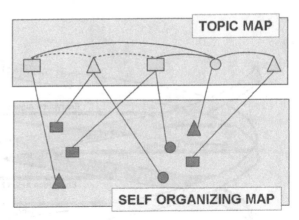

Fig. 6.9 Multiple I^{2}A Association framework

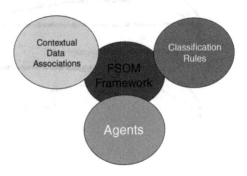

encryption/decryption scheme for every combination of subject, information, and context. Security levels for all three must match users' credential and need-to-know, in order for the information to be disseminated, or access granted. Figure 6.10 illustrates this concept.

As information is obtained, based on the user's request, the connections to other information and contexts of the information, as it relates to other information, is always expanding. It is necessary to create a continuously evolving, secure framework to carry the notion of information, subject, and context in a compact format that can accommodate multi-level security in one construct. The multi-dimensional fractal encryption framework will accommodate this.

For each combination of subject, context, and information, there is a separate fractal encryption algorithm, seeded by a 256 bit encryption key. Each combination equates to one of the quantum fractal eigenstates and the "information fractal" can be unpacked at any security level and need-to-know. This allows a

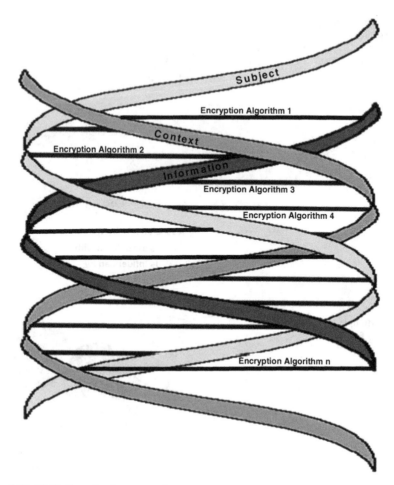

Fig. 6.10 Multi-dimensional quantum fractal encryption triple helix

continuously evolving, secure, multi-level security, Contextual Need-to-Know/
Need-to-Share framework for multi-level security system.

1. Reduction in data acquisition and recognition time.
2. Improved efficiency for autonomous decision support.
3. Improved processing and reporting timeliness.
4. Improved decision support quality.
5. Effective knowledge and decision management.

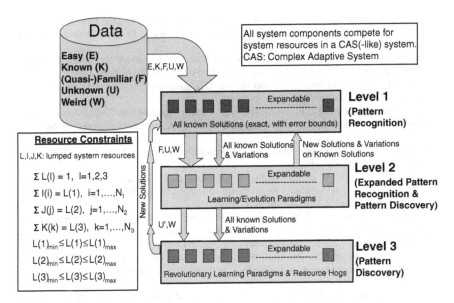

Fig. 6.11 I²A adaptation architecture

Conclusions and Discussion

We believe this framework provides an architecture and methodology for protection and authorization of classified information within real-time environments and situations. The use of the I²A architecture, combined with the topical maps and encryption schemes described have the potential to radically change and enhance secure environments in the future. With this, combined with the adaptation framework shown in Fig. 6.11, the system can learn, self-adapt, and react to rapid changes in situational conditions, while still providing a very secure framework for classified data/access. More work is needed to refine the agent technologies and learning sets, but we feel this has much potential.

References

1. Crowder, J. A. "Adaptive Learning Algorithms for Functional Constraints on an Evolving Neural Network System." NSA Technical Paper CON_0013_2002_003 (2002).
2. Crowder, J. A., "Integrating an Expert System into a Neural Network with Genetic Programming for Process Planning." NSA Technical Paper TIT_01_01_013_2001_001 (2001).
3. Crowder, J. A., "Machine Learning: Intuition (Concept) Learning in Hybrid Genetic/Fuzzy/Neural Systems." NSA Technical Paper CON_0013_2003_009 (2003).
4. Crowder, J., Barth, T., and Rouch, R., "Evolutionary Neural Infrastructure with Genetic Memory Algorithms: ENIGMA Theory Development." NSA Technical Paper, Denver (1999) ENIGMA_1999_004.

5. Crowder, J., Barth, T., and Rouch, R., "Learning Algorithms for Stochastically Driven Fuzzy, Genetic Neural Networks." NSA Technical Paper, Denver (1999) ENIGMA_1999_002.
6. Crowder, J., Barth, T., and Rouch, R., "Neural Associative Memory for Storing Complex Memory Patterns." NSA Technical Paper, Denver (1999) ENIGMA_1999_003.
7. Jacobs, R., Jordan, M., Nowlan, S, and Hinton, G., "Adaptive Mixtures of Local Experts." Neural Computation, Vol. 3 (1991).
8. Young-Woo, S., Giampapa, J., and Sycara, K., "A Multi-Agent System for Enforcing "Need-to-Know" Security Policies." International Journal of Cooperative Information Systems (2004).
9. K. Sycara. Negotiation planning: An AI approach. European Journal of Operational Research, 46:216–234, 1990.
10. Y. Arens, C. Y. Chee, C.-N. Hsu, and C. A. Knoblock. Retrieving and integrating data from multiple information sources. International Journal of Intelligent and Cooperative Information Systems, 2(2):127–158, June 1993.
11. F. Brazier, B. D. Keplicz, N. R. Jennings, and J. Treur. Formal specification of multi-agent systems: a real-world case. In First International Conference on Multi-Agent Systems (ICMAS'95), pages 25–32, San Francisco, CA., June 12–14 1995.
12. Chandramouli, R. and Sandhu, R., "Role based access control features in commercial database management systems." Proceedings of 21st National Information Systems Security (1998).
13. Giuri, L. and Iglio, P., "Role Templates for Content-Based Access Control." Proceedings of ACM Workship on Role Based Access Control, pp. 153–159 (1997).

Chapter 7
Cyber-physical Systems Security

Md E. Karim and Vir V. Phoha

Introduction

Concerns with the security of the cyber-physical systems include the malicious attempts by an adversary to intercept, disrupt, defect or fail cyber-physical systems that may affect a large group of population, an important government agency or an influential business entity by denying availability of services, stealing sensitive data, or causing various types of damages, as well as the security breaches in small scale cyber-physical systems that may affect few individuals or relatively smaller entities [1, 2].

Large scale cyber-physical systems are vulnerable to physical attacks due to their wide exposures usually over a large geographic area. They are vulnerable to cyber attacks because of their network based accessibility that allows exploitation of systems vulnerabilities remotely. The integration of cyber and physical components in a cyber-physical system introduces another category of vulnerabilities that involves interception, replacement or removal of information from the communication channels. Thus, as shown in Fig. 7.1, the vulnerability space in a cyber-physical system includes a physical, a cyber and an integration component. In this paper we briefly describe the security issues, associated challenges and possible measures for each of these components.

M. E. Karim · V. V. Phoha (✉)
Center for Secure Cyberspace, Louisiana Tech University, Ruston, LA 71272, USA
e-mail: phoha@latech.edu

M. E. Karim
e-mail: mdekarim@latech.edu

S. C. Suh et al. (eds.), *Applied Cyber-Physical Systems*,
DOI: 10.1007/978-1-4614-7336-7_7,
© Springer Science+Business Media New York 2014

Fig. 7.1 Vulnerability space in a cyber-physical system

Physical Component

Large scale cyber-physical systems often involve physical infrastructures such as flow networks [3], and numerous data origination points. Sensors are spread over those points to capture the data generated. Collected data are then forwarded via networks to one or more central locations known as sinks or base stations. Those data are analyzed and appropriate responses are made, either locally at the sink or at a remote system.

Physical infrastructures such as pipelines are one of the weakest security links in a cyber-physical system. It is practically not possible, for most of the large scale cyber-physical systems, to protect their geographically dispersed physical infra-structures from vandalism. An adversary can damage an electric gridline, remove railway tracks or inject cyanide in a waterline. Each of these examples can have very serious consequences. Sensors designed to detect the indicators of possible violations of physical infrastructures can assist in identifying such vandalisms and minimize associated damages. This is particularly significant for large scale physical systems where immediate detection of vandalism is near impossible otherwise. However, the sensors, whether deployed to monitor vandalisms of physical infrastructure or to collect data from other critical data origination points, are themselves vulnerable to vandalism.

Sensor networks consist of many small components each of which is subject to physical capture. An adversary can remove or destroy the sensors from the field creating a coverage hole, as shown in Fig. 7.2, and disrupting transmission of critical data. It can also corrupt or replace sensors and inject erroneous data into the system and fail the decision making system that depends on those data.

Various schemes, primarily graph theoretical and anomaly detection based, have been proposed for detecting coverage holes or identifying compromised sensor nodes that can detect the absence, corruption or replay of sensor data leading to the detection of possible vandalisms. If sensor networks can withstand the attacks against the data encryption and replay prevention schemes then it should be difficult for an adversary to vandalize a sensor without an anomaly being noticed at the recipient end. Schemes for the detection of vandalisms in a sensor networks are not matured yet. It is expected that the effective automated moni-toring of physical infrastructures as well as the sensor networks by identifying anomalies in the sensed input will be possible in the near future.

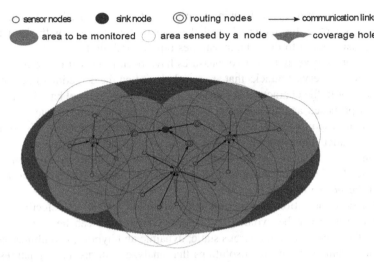

Fig. 7.2 The area *shaded gray* shows the coverage holes due to missing sensors

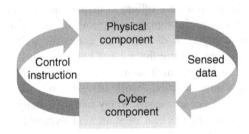

Fig. 7.3 The control loop in cyber-physical system

Cyber Component

Cyber component provides computational and control supports to cyber-physical systems (Fig. 7.3). It facilitates the fusion and analysis of data received from various sources and the overall decision making process. Remote network access that facilitates efficient interaction among various, possibly physically isolated, collaborating units of a cyber-physical system as well as efficient system administration, is an integral part of the cyber component. Such accessibility, however, also opens the door to the adversary for launching cyber attacks. These attacks may include: the denial of service, information corruption, destruction and exfiltration, and defective operation of the systems.

Denial of service attacks occur when an adversary creates an artificial mechanism, such as generation of frequent requests from a compromised distributed network, to keep the computing resources in the targeted system unnecessarily busy delaying or denying services to the legitimate requests [4]. Denial of service attacks are common in cyber domain; a variation of them, referred to as denial of sleep attacks targets the battery life of the sensors by engaging with them frequently causing them to drain their batteries rapidly and die [5].

Various preventive and reactive measures have been proposed in the literature against denial of service attacks that are mostly based on the detection of deviation in the usage or traffic characteristics expected in a system. Protocol based solutions have been proposed against denial of sleep attacks for the sensor nodes by minimizing their responsiveness to random requests; a common approach is to designating a leader node in a cluster that interacts with the world external to that cluster and letting the rest of the nodes in that cluster periodically communicate with the leader. Different nodes in a cluster take turn as the leader so that a not does not die before the others.

Information corruption, destruction and exfiltration as well as defective operation of a system can be avoided if a system compromise can be detected and nullified. There have been numerous static, dynamic and hybrid (a combination of static and dynamic approaches) solutions that analyze patterns and signatures in program codes and behavior of the program executions and identify the presence of malicious agents in the systems and help the system administrator to disable them [6–9]. The static solutions are offline in nature and they include computing entropy or looking for specific signatures in codes, comparing programs with known malwares and analysis of disassembled codes for activity-space exploration. The dynamic solutions analyze the runtime behavior, such as system calls, of a program and block its execution if some suspicious sequence of operations is committed. While dynamic solutions work in real time, they cannot detect a malware if its behavior is not expressed. Many malware analysts prefer a hybrid solution to take advantage of the best of static and dynamic analyses. Malware authors obfuscate the code as well as the behavior of malware to defeat malware analysis. Deobfuscation of malware is one of the major primary concerns today that has been addressed through mathematical modeling, code transformation and normalization, and weighing on game-theoretical approaches by making obfuscation choices difficult for the adversaries instead of finding a full proof solution for them.

Different approaches have been proposed to detect intrusion in cyber systems, both at network level and host level. They operate either by an anomaly in network activity such as of bandwidth, ports and protocols, or comparing network flows with pre-determined attack signatures and may suffer from delays from the onset of the attacks to their detection. In real time cyber-physical systems, that have been increasingly using embedded systems, many intrusions can be detected through static timing analysis [10].

Unauthenticated access to a system by a disgruntled insider or an adversary having a stolen password is another dominant issue in cyber security. Behavioral biometric based solutions that analyze, say, a user's typing dynamics, mouse maneuvers or computer usage patterns have been proven to be effective against such attacks, although there have been some concerns whether these solutions can withstand sophisticated spoofing [11].

If outbound flow of information from a system that risks being compromised is forced to pass through a secure system, we can host a monitor in that secure system to detect and prevent information exfiltration. Some of the major approaches noticed in the literature to prevent information exfiltration include: (i) statistical testing based methods, (ii) keystroke/mouse-click association based methods, (iii) packet marking based methods, (iv) heuristic rule based filtering and (v) blacklist based egress filtering. The first four of these approaches largely remain vulnerable to mimic attacks and the blacklist based ones suffer from the requirement for frequent manual intervention and cannot guarantee sufficient completeness of the blacklist. Statistical testing based methods observe different statistical properties of malicious and benign traffic and train a classifier or an ensemble of classifiers for the future classification of unknown traffic [12]. Attributes most frequently used in statistical testing include: header signature, new connection establishment rate, packet size, upload/download bandwidth, ARP request rate, ICMP echo reply rate, request regularity, request time of the day and packet structure. Keystroke/mouse-click association based methods correlate the timing of keyboard or mouse activity to the timing of outbound traffic [13]. Packet marking based methods mark all outgoing requests at the application level [14]. A remote entity receiving and forwarding requests to the destination verifies if the requests are marked before they are forwarded. Firewalls operate based on heuristic rule sets with 50 average number of rules although about 1 % firewalls have 1000 or more rules [15]. A review on their limitations is available in [16]. NuFW, a new generation of firewall, uses senders' profiles to mitigate attacks such as insider threats and may help in preventing information exfiltration [17].

An adversary can defective operation of a cyber-physical system if it can compromise the control loop. Several solutions have been proposed on graceful degradation (where the system continue operating under failure) and survivability under such attacks [18].

Integration

Security issues involving integration space are very specific to cyber-physical systems and any discussion on the security of cyber-physical systems primarily focus on these issues. The main security issues in integration space involve security of flow of information [19–21]. There are numerous ways an adversary can intercept and exploit the communication from the physical sources to the sink (decision making unit) and vice versa.

An adversary can physically access or replace a node located at a critical point in a sensor network, or remotely update it with malicious code and take it over. It is difficult to locate and reset the compromised devices or reload the codes on them. In addition, responses to the failures in the subsystems with different ownership are difficult to coordinate.

○ sensor nodes   ● base station   ◎ aggregator nodes   ——→ communication links

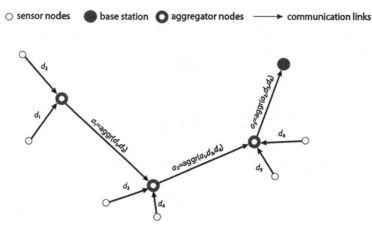

Fig. 7.4 Routing tree based data aggregation

Misleading information transmitted from the compromised nodes can lead to disastrous operation of cyber-physical systems. Many cyber-physical systems have real-time requirements and introduction of delay in the communication channel can result in cascading failures triggering one failure by the other [22–24]. Violation of confidentiality from the compromised nodes is another pressing issue that demands a satisfactory solution [25–27].

Since capture and compromise of a critical node in a sensor network is relatively easy, making a sensor network resilient under a compromise is the ideal solution that researchers are primarily seeking for.

Redundancy of the sensors is a common sense solution to many of the commonly encountered failures of sensors. However, in a sensor network, not all of the sensor nodes are equally important in terms of the value of information they contain and level of influence they have on the overall operation.

As shown in Fig. 7.4. Routing tree based data aggregation, in a typical sensor network based system data from different sources are aggregated at some intermediate nodes, known as data aggregators, on the way to the sink to filter out redundant data so that they get routed as fewer flows. This helps minimize the number of transmissions required to send the data to the sink, which in turn reduces the total energy spent by the network during the data transfers. In addition, it reduces the amount of bandwidth required for the data transfers.

As a consequence of the above schemes sensor located at a leaf node of the routing tree may have the least influence while a sensor located up above the routing tree and performing aggregation may be of high value. If an adversary can recognize high value sensors, it can selectively attack them without being overwhelmed by the abundance of the sensors. Being able to hide sensor network topology and routing infrastructure can make it difficult for an adversary to steal valuable information as well as corrupt massive amount of information [28]. For instance, the lesser association can be established between a terminal node and an aggregator node, the better we can limit the opportunities for compromising the aggregator node (by making it difficult to recognize).

Schemes for hiding topology may make the job for an adversary difficult but they do not guaranty resilient operation of a sensor network. It is important that we are able to identify a node that is compromised [29, 30]. It is also important to adopt some encryption based trust management scheme that can identify with high confidence which of the sensor nodes may or may not be trusted. However, in a sensor network end-to-end encryption from the leaf nodes to the central base station is not possible because, as mentioned before, many nodes located in between aggregate data from the lower level of the topology that cannot be done with encrypted data. Lightweight key based encryption schemes that can assist trust management as well as support data aggregation is a challenge towards achieving resilience in sensor networks.

Conclusion

Security aspects and associated models for cyber-physical systems vary from systems to systems and researchers envision their unified future differently. The common objective is to make sure that the availability, integrity and the confidentiality of cyber-physical systems are maintained are under an attack by resisting the attack or recovering from it or, through graceful degradation [18]. Determining generic security policies for cyber-physical systems is an important first step towards achieving those objectives under different criteria [31]. Security-aware platforms [32] and protocols [33] as well as devices (such as secure cyber-physical couplings, resulted from the new advancements in semiconductor technologies [34]), designed with cyber-physical systems in mind are eventually going to determine how the security approaches to cyber-physical systems evolve in the future.

References

1. N. Adam, "Cyber-physical systems security," presented at the Proceedings of the 5th Annual Workshop on Cyber Security and Information Intelligence Research: Cyber Security and Information Intelligence Challenges and Strategies, Oak Ridge, Tennessee, 2009.
2. E. K. Wang, et al., "Security Issues and Challenges for Cyber-Physical System," presented at the Proceedings of the 2010 IEEE/ACM Int'l Conference on Green Computing and Communications \& Int'l Conference on Cyber, Physical and Social Computing, 2010.
3. M. Yilin, et al., "Cyber-Physical Security of a Smart Grid Infrastructure," *Proceedings of the IEEE,* vol. 100, pp. 195–209, 2012.
4. J. Mirkovic, et al., *Internet denial of service: attack and defense mechanisms*: Prentice Hall, 2005.

5. M. Brownfield, et al., "Wireless sensor network denial of sleep attack," in *Information Assurance Workshop, 2005. IAW '05. Proceedings from the Sixth Annual IEEE SMC*, 2005, pp. 356–364.
6. A. Dinaburg, et al., "Ether: malware analysis via hardware virtualization extensions," presented at the Proceedings of the 15th ACM conference on Computer and communications security, Alexandria, Virginia, USA, 2008.
7. M. I. Sharif, et al., "Impeding Malware Analysis Using Conditional Code Obfuscation," in *NDSS'08*, 2008.
8. C. Willems, et al., "Toward Automated Dynamic Malware Analysis Using CWSandbox," *Security & Privacy, IEEE*, vol. 5, pp. 32–39, 2007.
9. A. Moser, et al., "Exploring Multiple Execution Paths for Malware Analysis," in *Security and Privacy, 2007. SP '07. IEEE Symposium on*, 2007, pp. 231–245.
10. C. Zimmer, et al., "Time-based intrusion detection in cyber-physical systems," presented at the Proceedings of the 1st ACM/IEEE International Conference on Cyber-Physical Systems, Stockholm, Sweden, 2010.
11. R. Chow, et al., "Enhancing cyber-physical security through data patterns," in *Proceedings of the Workshop on Foundations of Dependable and Secure Cyber-Physical Systems*, 2011.
12. B. Thuraisingham, "Data mining for security applications: Mining concept-drifting data streams to detect peer to peer botnet traffic," in *Intelligence and Security Informatics, 2008. ISI 2008. IEEE International Conference on*, 2008, pp. xxix–xxx.
13. R. Gummadi, et al., "Not-a-Bot: improving service availability in the face of botnet attacks," presented at the Proceedings of the 6th USENIX symposium on Networked systems design and implementation, Boston, Massachusetts, 2009.
14. K. Xu, et al., "Data-Provenance Verification For Secure Hosts," *IEEE Trans. Dependable Secur. Comput.*, vol. 9, pp. 173–183, 2012.
15. P. Gupta, "Algorithms for routing lookups and packet classification," PhD Thesis, Stanford University, Stanford, CA, USA, 2000.
16. A. X. Liu and M. G. Gouda, "Diverse Firewall Design," *Parallel and Distributed Systems, IEEE Transactions on*, vol. 19, pp. 1237–1251, 2008.
17. N. C. Team, "NuFW firewall: Now User Filtering Works," 2008.
18. A. A. Cardenas, et al., "Secure Control: Towards Survivable Cyber-Physical Systems," in *Distributed Computing Systems Workshops, 2008. ICDCS '08. 28th International Conference on*, 2008, pp. 495–500.
19. R. Akella, et al., "Analysis of information flow security in cyber–physical systems," *International Journal of Critical Infrastructure Protection*, vol. 3, pp. 157–173, 2010.
20. T. T. Gamage, et al., "Enforcing Information Flow Security Properties in Cyber-Physical Systems: A Generalized Framework Based on Compensation," presented at the Proceedings of the 2010 IEEE 34th Annual Computer Software and Applications Conference Workshops, 2010.
21. T. Gamage, et al., "Information flow security in cyber-physical systems," presented at the Proceedings of the Seventh Annual Workshop on Cyber Security and Information Intelligence Research, Oak Ridge, Tennessee, 2011.
22. C. Neuman, "Challenges in Security for Cyber-Physical Systems," in *Workshop on Future Directions in Cyber-physical Systems Security*, 2009.
23. H. Tang and B. M. McMillin, "Security Property Violation in CPS through Timing," presented at the Proceedings of the 2008 The 28th International Conference on Distributed Computing Systems Workshops, 2008.
24. F. Mueller, "Challenges for Cyber-Physical Systems: Security, Timing Analysis and Soft Error Protection," in *Proc. of the National Workshop on High Confidence Software Platforms for Cyber-Physical Systems*, 2008.
25. T. T. Gamage, et al., "Confidentiality Preserving Security Properties for Cyber-Physical Systems," presented at the Proceedings of the 2011 IEEE 35th Annual Computer Software and Applications Conference, 2011.

26. T. Kohno, "Security for cyber-physical systems: case studies with medical devices, robots, and automobiles," presented at the Proceedings of the fifth ACM conference on Security and Privacy in Wireless and Mobile Networks, Tucson, Arizona, USA, 2012.

27. R. Mitchell and I.-R. Chen, "Behavior Rule Based Intrusion Detection for Supporting Secure Medical Cyber Physical Systems," in *Computer Communications and Networks (ICCCN), 2012 21st International Conference on*, 2012, pp. 1–7.

28. Z. Quanyan, et al., "A hierarchical security architecture for cyber-physical systems," in *Resilient Control Systems (ISRCS), 2011 4th International Symposium on*, 2011, pp. 15–20.

29. M. Mathews, et al., "Detecting Compromised Nodes in Wireless Sensor Networks," in *Software Engineering, Artificial Intelligence, Networking, and Parallel/Distributed Computing, 2007. SNPD 2007. Eighth ACIS International Conference on*, 2007, pp. 273–278.

30. P. R. Nalabolu, "Detecting Malicious Code in Sensor Network Applications Using Petri Nets," M.S., Oklahoma State University, Oklahoma City, OK, USA, 2007.

31. K. K. Fletcher and L. Xiaoqing, "Security Requirements Analysis, Specification, Prioritization and Policy Development in Cyber-Physical Systems," in *Secure Software Integration & Reliability Improvement Companion (SSIRI-C), 2011 5th International Conference on*, 2011, pp. 106–113.

32. M. Azab and M. Eltoweissy, "Defense as a service cloud for Cyber-Physical Systems," in *Collaborative Computing: Networking, Applications and Worksharing (CollaborateCom), 2011 7th International Conference on*, 2011, pp. 392–401.

33. G. S. Lee and B. Thuraisingham, "Cyber-physical systems security applied to telesurgical robotics," *Comput. Stand. Interfaces*, vol. 34, pp. 225–229, 2012.

34. O. Al Ibrahim and S. Nair, "Cyber-physical security using system-level PUFs," in *Wireless Communications and Mobile Computing Conference (IWCMC), 2011 7th International*, 2011, pp. 1672–1676.

Chapter 8
Axiomatic Design Theory for Cyber-Physical System

Cengiz Togay

Introduction

Cyber-physical systems (CPSs) include hardware, software, and network entities, while considering the limitations of human element as well. CPSs are based on the embedded systems. In situations requiring reliable operations, embedded system designs are generally preferred. Embedded systems generally provide more reliability than general-purpose computing [1]. A key difference between a pure embedded system and CPS is the existence of networking. Therefore, CPSs can generally be defined as systems of systems. Alternatively, CPSs can also be considered as a component based/oriented systems [1, 2]. Heterogenous components (both software and hardware) integrated through a network are depicted in Fig. 8.1. We can even represent network as a component since there is different time-aware network technologies with different communication interfaces such as FlexRay [3] for automative domain. However, component oriented software engineering methodologies are more suitable for soft components than physical components because of their nature [1, 4–7]. In CPS, there is a close communication with surrounding physical environment. Therefore, in contrast of well-defined, secured, and synchronized software platforms, we should consider designing a multidisciplinary, concurrent, and distributed system. In CPS, cyber components change the physical world through physical entities. Therefore, complexity of the system is increased because of unpredictable nature of world. There are various challenges in implementing a CPS. Because of networked embedded systems and cyber units, by definition, CPSs are concurrent systems. When considering numerous unpredictable parameters associated with the environment especially the concurrency issues, CPSs can be defined as multidisciplinary complex systems as well. We should consider various parameters during design. For example, while we are considering design of software components, we

C. Togay (✉)
Netas, Netas-Alemdag Caddesi 171 34768 Umraniye, Istanbul, Turkey
e-mail: ctogay@netas.com.tr

S. C. Suh et al. (eds.), *Applied Cyber-Physical Systems*,
DOI: 10.1007/978-1-4614-7336-7_8,
© Springer Science+Business Media New York 2014

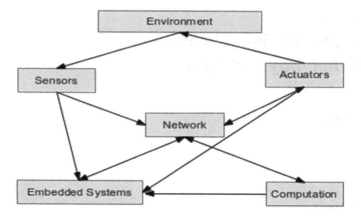

Fig. 8.1 Cyber-physical systems

should consider physical world parameters such as time, memory, and network
throughput as well as the nature of physical entities such as actuators' response
time. Like all systems, phases of CPSs are also design and implementation. In
implementation phase, there are some technical challenges listed in the following
section, for instance; time awareness (networks, operating systems, and software
entities). There are many engineering methods and architectures applied to CPS
some of them as shown below:

- An extended Service-Oriented Architecture (SOA) is applied to cyber-physical
  systems [7–9]. Dynamic service composition requires ontological information
  about services provided and their use.
- In Aspect Oriented Model Driven Architecture [10, 11]: Time constraint is
  considered as an aspect.
- Modeling and Analysis of Real-Time and Embedded Systems (MARTE) [12]
  and OMG Systems Modeling Language (SysML) [13] are utilized to design
  multi-disciplinary systems. Although MARTE and SysML have various views
  to support embedded systems, they are originally defined for software development [14]. Therefore, some new UML items especially for physical entities
  such as time constraints are added and integrated with MARTE and SysML
  [14]. SysML based on UML defines syntax of the diagrams but generally suffers
  from lacks of semantic expressiveness [15].
- CPS systems through composition of component mechanisms are studied in
  [16]. Composition is considered in two levels namely in design time and in runtime. In the run-time, composition approach should allow dynamic binding. For
  example, in SOA whenever a service is made available for use, another service
  can effectively use it as needed.
- As part of a CPS, wireless sensor networks, REMORA event-driven programming is proposed [17]. REMORA provides XML based component interface but
  does not provide the concurrency, task scheduling, and interrupts handling.

- In most of the CPSs, timing has critical importance. IEEE 1588 protocol is introduced to enabling system wide precise clock for heterogeneous systems with an accuracy less than tens of nanoseconds over hundreds of meters stretched networks [18, 19].
- Because of heterogeneity, time can be represented as time-stamped events. Therefore, developers can specify interactions among distributed components without considering low-level details. One example of such systems are High Level Architectures (HLA) and its implementation Runtime Infrastructures (RTI) [20, 21]. HLA defines standards for general-purpose distributed computer simulations. Another example is Programming Temporarily Integrated Distributed Embedded Systems (PTIDES) which relies on network time synchronization [19]. In both RTI and PTIDES, time-stamped messages are delivered to components.
- Ptolemy projects provide open source environments for modeling, simulating, and designing concurrent, real-time, and embedded systems [22].
- FlexRay, an automative network communication protocol, is developed for high performance components. All components (engine, transmission, braking, steering, and suspension) are connected to network. FlexRay provides effective response time (less than 100 microseconds) for "real-time" operations. Each component has capability to get all messages in the bus. In FlexRay Time Division Multiple Access (TDMA) schema is used. Because of TDMA, consistency of the data deliver is provided.

In this chapter, we concentrate on the design phase and proposing the utilization of Axiomatic Design Theory (ADT) [23]. Design is defined as mapping process between what we want to achieve and how we want to implement in [23]. ADT is an approach supporting the design of complex multidisciplinary systems. There are many disciplines using axiomatic design theory including mechanical engineering, aerospace engineering, civil engineering, and software engineering. Furthermore, we review the key aspects of ADT suitable for designing complex CPSs. For consistency, we are using in our exposition component oriented approach [4]. It has been demonstrated that component oriented approach can handle interactions among software and hardware modules among others. For analysis and design purposes, these complex systems can be decomposed to modules through ADT as well. Dependencies among components specify the system complexity in terms of security, durability, and maintenance. Therefore, during the cyber-physical systems' design, dependency should be considered. One of the axioms of ADT is independence axiom, which is utilized during system decomposition. Loosely coupled components' composition and flexible binding processes are much easier than tightly coupled ones. In our previous studies, we have used information axiom of ADT to measure congruity of the components [24]. Concurrent system design and implementation is generally a highly complex process. Therefore, design approaches for CPSs should provide comprehensive support.

This paper aims to analyze the use of Axiomatic Design Theory to design CPSs. A brief introduction is given in section two about CPS and in section three about ADT. The discussion section will be section four.

Cyber-Physical Systems

Cyber-physical systems (CPSs) present a holistic approach for representation and modeling of complex systems. In such a system, entities (sensors, actuators, control processing units, and communication devices) are collaborated through a network [25]. Computation processes are connected to physical entities for monitoring and controlling physical processes [10]. CPSs are multidisciplinary, multi-technology distributed and heterogeneous systems due to their nature. There are various applications of CPSs: high confidence medical systems, health-care service (observe patients' condition, analyze context information, communicate with doctor or nurse), traffic control and safety, advanced automotive vehicle systems (collision avoidance), autonomous systems (search and rescue, firefighting, exploration), process control, energy conservation, electric power grids (power electronics, power grid, and embedded control software), cooling systems of nuclear plant, environmental control, avionics, instrumentation, critical infra-structure control, distributed robotics (telepresence, telemedicine), defense systems, manufacturing, unmanned vehicle, intelligent transportation systems include cyber aspects (wireless communication and computer based control) and physical aspects (movement in space and real time interfacing with environment), sonar systems, tour guidance, and smart structures (improve energy efficiency) [1, 2, 10, 14, 16, 26–30]. Because of nature of the physical environment, it is not predict-able. Therefore there is a need to adaptable CPS for subsystem failures. As an engineering challenge, it is very hard to design and develop such systems. Adaptability is a key feature of such problems. For instance, GPS navigators are prepared to reroute if a need arises. They sense the out of route situation through GPS values and initiate reroute. Component oriented systems are based on hier-archical structure. In each level, entities assume that a lower level provides reli-ability. For instance, during design and development of a web service we do not consider TCP/IP protocols. We assume that networking is provided by lower level layers. Another example, we can write a simple C code without concurrency, it is expected that will work the same way for the same environment. However, during the execution of the code numerous errors can occur, while they can be solved through various methods.

Physical entity producers declare the tested ranges for the reliability of their product. Physical systems through composition of the physical and cyber entities require stringent overall testing environments. Any upgrade or error correction procedures also requires overall re-testing. Because of testing costs CPS manu-factures stockpile all components and avoid the upgrades (i.e. aircraft manufac-turers) [1]. Any upgrade can change/affect various parameters including time constraints [1]. In hierarchical structures, problems and some features are hidden or taken cared by low levels. For instance if a packet is not received correctly it can be retransmitted without any clue to higher levels. Another example is signal strength, which can be valuable for higher layers. Still another feature is time concept; it is generally hidden by higher layers because of abstractions [1]. Some tasks should be done in a specific time, however with todays' technologies; we can only weakly determine worst case execution time (WCET) bounds through bench tests and with modern processors, which include caches, and dynamic dispatch among others [1]. It should be noted that synchronous embedded systems generally behave deterministically in terms of time. In real time systems, this is very hard to determine because of concurrency. We have identified following properties for cyber-physical systems [1, 7, 15, 16, 24, 25, 26, 30, 31]:

- *Tightly Integrated*. Computation and physical entities are closely integrated. In CPS, physical entities are monitored and controlled by computation entities.
- *Networked*. Entities are connected to each other through heterogeneous network (such as wired/wireless network, WLAN, Bluetooth, and GSM).
- *Adaptability*. Because of the unreliable network and unstable physical environment, entities in CPS should have capability to reconfigure.
- *Automation*. Although human interaction part of the CPS, entities should have automation capability [2, 16, 31]. In unexpected situations, CPSs should adapt itself to the new situation.
- *Non-functional requirements*. Due to security and reliability requirements especially for critical systems, certification can be applied. In some applications, because of constraints such as time and resource limits, entities should specify their interaction with environment.
- *Distributed*. All resources and entities are spread on the environment and interacted through network.
- *Multidisciplinary*. Entities form CPS can include different kinds of engineering approaches and methodologies.
- *Limited resources*. Although, todays resources are beyond of the requirement for most applications, CPSs should consider resource limit and constraints.
- *Time awareness*. In general-purpose software, time is considered as performance instead of correctness [15]. Most of the CPSs are time critical systems. Distributed cyber-physical components have to succeed their tasks in specific time range, include all computations and communications start with sensors and end with actuators, depending on applications.
- *Dependability*. CPSs should be dependable (reliable, maintainable, available, safe, and secure) since service failures should be less than acceptable ranges. Fault management approaches can be applied [24].
- *Predictability/Determinism*. Each entity in the cyber-physical system should provide services in predictable way in terms of time, used resources, and effect to environment. Physical systems are not deterministic systems. There is a chance that executing same model with same input can produce different output than expected [15].
- *Risk of casualties*. As given CPS examples include healthcare systems and cooling systems of nuclear plant, error in CPSs might result with serious casualties.

There are many challenges such as high safety requirements, distributed structure, problems of real time nature, heterogeneity, unreliable networks, mobility, tight environmental coupling, and validation and verification [1, 8, 16, 25–30, 32–34]:

- *High safety requirements.* CPS systems such as a tele-operated surgery or collision avoidance systems should provide the safety requirement otherwise CPS can cause loss of human life or catastrophic damages (an error in cooling system of a nuclear plant). Therefore, each entity in the system should be reliable that means system should be protected against random failures. System should also be adaptable and can heal itself. For instance, health monitoring and management systems should react faults in real-time [28],
- *Distributed structure.* Distributed structure affects to reliability of the CPS. There are various impacts due to the nature of distributed structure such as communication problems and scalability.
- *Problems of real time nature.* Concurrency and asynchronous communication are one of the important challenges during design and testing.
- *Heterogeneity.* CPS systems are composed of various technologies and devices. Such as, sensor nodes, mobile devices, computers, embedded systems, different types of communication protocols (such as Bluetooth, ZigBee, and 802.11) Due to the constraints such as battery life, different physical entities require different technologies, for instance communication, memory, and processor.
- *Unreliable Networking.* Both environmental factors (unexpected noise sources) and energy consumption constraints affects to reliability of the network. These kinds of affects cannot be considered during design but constraints can be defined. Especially for time critical systems such as health monitor or energy transportation problems, communication constraints (time delays and throughput) are getting interest [30]. It can decrease performance and even cause catastrophic results. Collaborative wireless sensors can be used to sense environment [33]. Wireless sensors provide advantage because of their self-healing property but at the same time dynamically routed packages cause predictability problem.
- *Mobility.* Mobility of physical entities effects the communication and processes. Communication devices have some range constraints and they should be considered in design. This kind of constraints increases the complexity of design. Mobility also affects networking. In CPS networks should have capability for self recovery.
- *Tight environmental coupling.* Utilized physical entities can be affected from environment. Such as crystal oscillators and capacitors are affected from temperature values. For these kinds of problems, CPS can be designed for all possible environment situations or CPS can adopt itself for a chancing environment.
- *Validation and verification.* Validation and verification of CPS is not an easy task because of the concurrency, out sourced development, cost, software created without formal models, or complex models where the verification process is inconvenient [16, 32]. Model checking is one method to check all possible states in terms of undesired states of system. However, these kinds of approaches are

developed for deterministic systems. CPS requires more attention to validation and verification of systems functional and non-functional properties of system because of tight interaction with environment and concurrent features. Because of interaction with physical world, each interaction can cause unexpected consequences. Therefore, during the evaluation of the composed system, affects to environment should be also considered in terms of goals.

- *Security*. Distributed and heterogeneous entities should be secured during their utilization. Entities should have operational goals such as closed-loop stability, safety, and liveness [25]. One challenge of the security is protection of each entity from outside affects in terms of these goals [25, 34]. Also other traditional security goals (integrity, availability, and confidentiality) need to be considered [25]. Sand box method can be used to provide some critical resources in terms of safety [32].
- *Testing*. CPSs are based on the previous embedded systems. However, embedded systems that are designed as closed box are differs from CPS due to lack of network capability [1]. Bench tests are not suitable for networked embedded systems due to lack of capability of all possible conditions [1]. Compositional verification and testing methods should be applied [26].
- *Lack of time awareness*. One of the important features of the CPS is time. To meet the needs of cyber-physical systems, operating systems and middle-wares must be evolved [35]. Generally, produced and improved operating systems and programming languages are solution for time irrelevant problems [1]. Time-stamp based event approaches can be used. Although, time synchronization through network concept is improved as mentioned before, still there can be problem because of the environment. Unexpected delays on synchronization can be problem.
- *Robustness and predictability*. Embedded systems design and development approaches are enhanced from efficiency to robustness and predictability [1]. Today's embedded hardware includes more memory and process capability than decades ago. Therefore, efficiency is not a problem for today's and tomorrow's technologies. However, we still cannot discard efficiency, for instance health monitor and management systems should decrease required time to identify faults [28]. Many composed system technologies such as aircrafts composed from various embedded systems preserve same components until its production stop because of certification and validation costs [1].
- *Abstractions*. Hierarchical structure is inevitable feature of complex systems. However, challenge is which information such as time should be hidden from upper layers? Also some techniques such as caches, dynamic dispatching is a problem for predictability of execution time. Therefore, designer should affect to operating systems (set up priorities and timer interrupts) [1].
- *Component search*. During service search to satisfy a specific task/goal, all related services should be investigated. Large numbers of physical and computational entities is a challenging problem [7].

Axiomatic Design Theory

Axiomatic Design Theory (ADT) introduced by Nam P. Suh is a methodology to consistently decompose systems in a top-down fashion [23, 36]. It is a multidisciplinary approach supporting design strategies in various engineering disciplines including software and mechanical engineering. There are implementations of the ADT such as Acclaro DFSS [37] general purpose tool and ADCO [5] component oriented design tool. ADT introduces four abstract concepts namely domains, hierarchies, zigzagging, and axioms. In domain concept, ADT proposes the consideration of problems in four interconnected domains which are customer, functional, physical, and process domains which is depicted in Fig. 8.2. In customer domain, customer requirements are identified. There is no explicit way to identify customer needs but various studies [5, 38] are proposed for ADT. Depending on customer needs (CNs), minimum set of independent requirements are defined as functional requirements (FRs) in functional domain. FRs define the expected services and functionality from the system without considering any implementation level problem or solution. At the same time, for each FR at least one Design Parameter (DP) in physical domain and one Process Variable (PV) in process domain are specified. There is a "What" and "How" relation between domains. DPs are defined to satisfy FRs and they can be interface items such as property or method names, components, and some abstractions in terms of software. DPs are satisfied with PVs (cyber and physical entities). For instance, in terms of object oriented programming, customer attributes, objects, data, and machine codes are represented in customer domain, functional domain, physical domain, and process domain respectively [23].

Second concept of the ADT is hierarchies. Typically, humans prefer to decompose problems to simple parts, usually exhibiting a hierarchy, and solve them. In ADT, root FRs, DPs, and PVs represent abstractions of the problem and a solution representation.

Third concept is zigzagging. In this concept, decomposition is applied to all domains, except customer domain, simultaneously. When defined a new FR, a DP

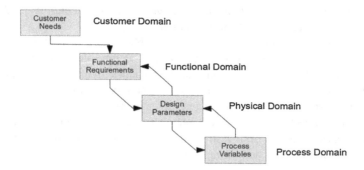

Fig. 8.2 Domains of axiomatic design theory (adapted from [23])

and a PV are also defined as depicted in Fig. 8.2. Decomposition continues to all FRs are satisfied by real cyber or physical entities.

Last concept of the ADT is axioms. Axioms by definition are generally accepted starting points that cannot be derived but they are accepted due to lack of counterexamples. Problems should be decomposed systematically based on axioms. Although there are various axioms in ADT, there exist two main axioms which are independence and information axioms. In the independence axiom, designer should conserve the independency of the FRs in terms of linked DPs or the independency of the DPs in terms of linked PVs. Interactions between sequential domains are represented in a design matrix as depicted in Fig. 8.3a and b. For instance, if Y is an FR, X can be DP or if Y is a DP, X can be PV. There are three kinds of design namely uncoupled, decoupled, and coupled designs depicted in Fig. 8.3c–e respectively. Module diagrams of the each design type are also represented in Fig. 8.3f. Module diagrams are used to represent system architecture. Depending on the independence axiom, only A_{11}, A_{22}, A33, and A44 should be 1 and others should be 0 as depicted in Fig. 8.3c. For FR-DP design matrix, each FR should be satisfied by exactly one DP. These kinds of designs are called uncoupled designs. FRs are satisfied without effecting each other. We can remove any FR from the system without effecting the overall system. In terms of implementation, if DP-PV matrix is uncoupled, entities of solution can be executed concurrently as depicted in Fig. 8.3f. If design matrix is triangular (all ones are located one side of the diagonal), then this kinds of designs are called as decoupled designs as depicted in Fig. 8.3d. Although, uncoupled designs are desired, decoupled designs are also acceptable. Dependencies between modules can be

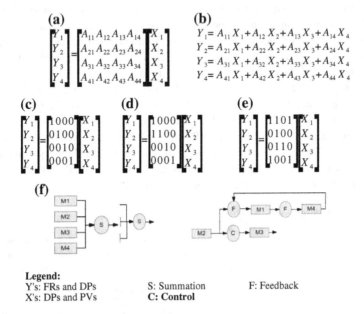

Fig. 8.3 Design matrices (adapted from [23])

seen in Fig. 8.3f. In last kinds of design called coupled design, ones are located in both sides of the diagonal as depicted in Fig. 8.3e. Coupled designs cause several effects on system such as maintenance and concurrency problems. As it can be seen in the module diagram of the coupled design in Fig. 8.3f, M4 requires M1 and M1 requires M4. If there is a time relationship between them, it is clear that a concurrency problem can occur. It should be noted that, some algorithms can be used to reallocate FRs and DPs to obtain uncoupled design if it is possible. Therefore, some designs can seem to be as coupled design although they are uncoupled, tool support is required to identify real coupled designs or reorganize matrix to decoupled design. When a FR is defined and designer cannot find any DP to satisfy it then the FR should be reconsidered. Although, ADT provides top to bottom decomposition, goal is design and implement a system. If any problem is figured out during design in a lower level, reevaluation process can be considered for overall system because of new FRs, DPs, PVs, and relationships among them.

Another axiom of the ADT is information axiom. Based on the information axiom, information content of the design should be minimized. Information content I_i represents probability P_i of satisfying FR_i and information content of entire system for all FRs is represented below.

$$I_i = -\log_2 P_i$$

$$I_{system} = -\sum_{i=1}^{n} (\log_2 P_i)$$

In complex designs, since probability of the FRs success low, information content of system is high. Probability of success depends on common area between what FRs require (design range) and system provide (system range) as depicted in Fig. 8.4. Probability of success is calculated common range over system range. For instance if you need a cooling device for food to keep between -50 and $-10\ °C$ and refrigerator can only cool between -30 and $10\ °C$, then probability of success of FR with given DP is 50 %. In terms of congruity between design and implemented components, information content can also be used [24]. Design algorithm based on ADT is listed in Table 8.1. Example design is depicted in Fig. 8.5.

In mature domains, all possible FRs and solutions can be represented in a design matrix. Subsets of the mature domain design matrix define special applications in a domain [4]. Mature domains can also be called flexible systems because of different sets of the FRs are used to satisfy system requirements in

Fig. 8.4 System, design, and common ranges of a functional requirement based on [23]

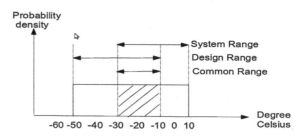

Table 8.1 System design based on Axiomatic Design Theory (adapted from [4] and [23])

Step	Description
1	Specify customer needs (feature models can be utilized)
2	Specify a functional requirement based on customer needs and previous functional requirements. Because of hierarchical structure of ADT, functional requirements should be specified most abstract at the root level and abstraction should be decrease in the lower levels
3	Specify design parameter to satisfy functional requirement. During specifying, design parameters as soon as possible we should map FRs to real components
4	Set dependencies in the FR-DP design matrix. All required DPs are marked in the FR-DP design matrix
5	Apply independence axioms to design matrix. Design should be uncoupled or decoupled
6	Specify process variable to satisfy design parameter
7	Set dependencies in the DP-PV design matrix. All required PVs are marked to satisfy DP in the DP-PV design matrix
8	Apply independence axiom to the DP-PV design matrix. Design should be uncoupled or decoupled
9	If all FRs are implemented, go to step ten otherwise go to step two
10	Compose components for the aimed product

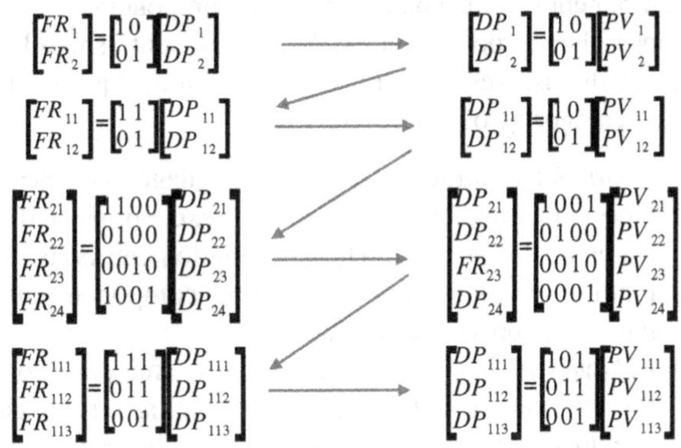

Fig. 8.5 Decomposition example with zigzagging process in ADT

$$\begin{bmatrix} FR_1 \\ FR_2 \end{bmatrix} = \begin{bmatrix} 1 & 0 \\ 0 & 1 \end{bmatrix} \begin{bmatrix} DP_1 \\ DP_2 \end{bmatrix}$$

$$\begin{bmatrix} DP_1 \\ DP_2 \end{bmatrix} = \begin{bmatrix} 1 & 0 \\ 0 & 1 \end{bmatrix} \begin{bmatrix} PV_1 \\ PV_2 \end{bmatrix}$$

$$\begin{bmatrix} FR_{11} \\ FR_{12} \end{bmatrix} = \begin{bmatrix} 1 & 1 \\ 0 & 1 \end{bmatrix} \begin{bmatrix} DP_{11} \\ DP_{12} \end{bmatrix}$$

$$\begin{bmatrix} DP_{11} \\ DP_{12} \end{bmatrix} = \begin{bmatrix} 1 & 0 \\ 0 & 1 \end{bmatrix} \begin{bmatrix} PV_{11} \\ PV_{12} \end{bmatrix}$$

$$\begin{bmatrix} FR_{21} \\ FR_{22} \\ FR_{23} \\ FR_{24} \end{bmatrix} = \begin{bmatrix} 1 & 1 & 0 & 0 \\ 0 & 1 & 0 & 0 \\ 0 & 0 & 1 & 0 \\ 1 & 0 & 0 & 1 \end{bmatrix} \begin{bmatrix} DP_{21} \\ DP_{22} \\ DP_{23} \\ DP_{24} \end{bmatrix}$$

$$\begin{bmatrix} DP_{21} \\ DP_{22} \\ FR_{23} \\ DP_{24} \end{bmatrix} = \begin{bmatrix} 1 & 0 & 0 & 1 \\ 0 & 1 & 0 & 0 \\ 0 & 0 & 1 & 0 \\ 0 & 0 & 0 & 1 \end{bmatrix} \begin{bmatrix} PV_{21} \\ PV_{22} \\ PV_{23} \\ PV_{24} \end{bmatrix}$$

$$\begin{bmatrix} FR_{111} \\ FR_{112} \\ FR_{113} \end{bmatrix} = \begin{bmatrix} 1 & 1 & 1 \\ 0 & 1 & 1 \\ 0 & 0 & 1 \end{bmatrix} \begin{bmatrix} DP_{111} \\ DP_{112} \\ DP_{113} \end{bmatrix}$$

$$\begin{bmatrix} DP_{111} \\ DP_{112} \\ DP_{113} \end{bmatrix} = \begin{bmatrix} 1 & 0 & 1 \\ 0 & 1 & 1 \\ 0 & 0 & 1 \end{bmatrix} \begin{bmatrix} PV_{111} \\ PV_{112} \\ PV_{113} \end{bmatrix}$$

Fig. 8.6 Example FR-DP design matrices

(a)

$$\begin{bmatrix} FR_1 \\ FR_2 \\ FR_3 \end{bmatrix} = \begin{bmatrix} 1 & 0 & 0 \\ 0 & 1 & 0 \\ 0 & 0 & 1 \end{bmatrix} \begin{bmatrix} DP_1 \\ DP_2 \\ DP_3 \end{bmatrix}$$

(b)

$$\begin{bmatrix} FR_{21} \\ FR_{22} \\ FR_{23} \end{bmatrix} = \begin{bmatrix} 1 & 1 & 0 \\ 0 & 1 & 0 \\ 0 & 0 & 1 \end{bmatrix} \begin{bmatrix} DP_{21} \\ DP_{22} \\ DP_{23} \end{bmatrix}$$

different times [4, 23]. There are also fixed FRs, DPs, and PVs where all system design should include them. ADT only considers functional requirements' independency but not the independency of physical entities. To identify dependencies of physical entities which are emerged after ADT based design, Design Structure Matrices can be used [39].

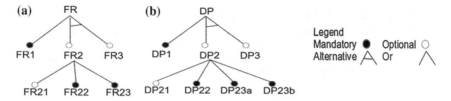

Fig. 8.7 FR and DP views based on feature model

Discussion

Cyber-physical systems (CPSs) include cyber and physical entities in a networking environment. Complexity of design and development process of a system requires systematic approaches. Axiomatic Design Theory (ADT) is an important candidate to design complex systems such CPSs. In ADT, problems are decomposed considering independency of the functional requirements and solutions (Design Parameters and Process Variables). Cyber-physical systems can be considered as integration/composition of components in a highly dynamic environment. We assume that these components are connected through a time-aware network. Every component can also be a composite component. A systematic component-oriented approach is proposed in [4, 5]. Components should provide enough information about themselves to entities which would require services from them. There are four categories of contracts [40]:

- *Syntactic.* Interface description languages are used to define the kinds of services, input and output parameters, exceptions, and some information about how we can call/use them. This information can be used in both static and dynamic binding situations. Since most of the components are provided as closed-box, we have only provided interfaces.
- *Behavioral.* Boolean assertions, pre and post conditions provide little more information for utilization of the specific service-call.
- *Synchronization.* Concurrency related information should be provided. Therefore, designer can handle synchronization issues during system design through some mechanism such as mutex.
- *Quality of service.* Some non-functional parameters such as maximum response delay, and average response time should be provided.

Since the lack of standard for hardware component interfaces, we assume that specific composed components such as Application Specific Integrated Circuits and networked sensors are designed with ADT and they have comprehensive interface. Mentioned categories define the minimum requirements. We need more ontological data about the services provided. Service Oriented Architecture is a candidate to provide more information about available services. However, design matrices as an interface provide more information such as the necessity of a service and systemic dependencies.

As mentioned before time is important for CPSs. Dependencies and junction points in design matrix for specific FRs and DPs provide the information about where we should be carefull during service-call. Also information axiom can be used. Information content can be used to measure time congruity between components. Complexity is related with uncertainty and information content represents uncertainty level of probability of success of FR [41]. Therefore, complexity of the system is quite related with information content of the system. Nonfunctional parameters can be evaluated through their information content. Time can be evaluated as one of the parameters of information content [41]. Other parameters related with design can also be evaluated. Therefore, complexity of the system can be measured. We can define exactly time ranges of FR and DP. For instance, if we need our robot move a specific distance with range tolerance and our component (engine) moves range of error because of environment, information content can be higher than our expectations. In this situation, we can change component or we get some precautions to handle problem such as using counter to measure what we get. It should be noted that our new proposed solution should be reevaluated in terms of new information content. For time parameter, we can also define composition constraints. For instance, we can expect from our components synchronized communication or asynchronous communication.

ADT provides a decomposition mechanism with axioms while helping to keep recording design decisions and identifying overall complexity items during CPS design. For instance, we can miss coupling in higher level for FRs, DPs and PVs. During decomposition, they can be identified. Therefore, ADT tools can be helpful to warn about this identified coupling situations.

System and component design can be considered as different phases [4, 5]. System design matrix can be an intersection of the component's design matrices. Each component's design matrix define capabilities of the component in terms of FRs and dependencies. System design matrix includes the DPs and PVs of the components with same or different FRs.

Applied technologies and components change over time. ADT separates requirement and implementation items through domain concept. In implementation level changes with improved technology only effects the DP and PV relationships. ADT also promotes independent designs. Therefore, ADT can be helpful to speed up validation and verification process.

Designs especially CPS designs should be flexible. However, instead of creating new design matrices for all alternative systems in a domain, all fixed or optional FRs, DPs and PVs can be represented in the same design matrix called mature domains [4, 5]. In [26], we have introduced new views to represent this kinds of relations based on Feature Model [42] representation. Therefore, which FRs, DPs, and PVs can be optional or mandatory and group relationships among them can be represented. FR and DP views of FR-DP design matrices in Fig. 8.6 are depicted in Fig. 8.7. As depicted in Fig. 8.7, FR1 is a mandatory FR, therefore without it design cannot be a solution. Such design elements are utilized to represent fixed parts of the system. There is alternative relationship between FR2 and FR3. Our CPS design can include either FR2 or FR3. If FR2 is selected as solution of the design then FR22 and FR23 has to be part of the solution but FR21 is optional. We can define different DPs to satisfy same FR depending on the non-functional requirements. For instance, there are two alternatives of the DP23 namely DP23a and DP23b. Also, some constraints (requires or excludes) can be defined between design items [5]. For instance, constraint "FR3 exludes DP23b" declares that if FR3 is selected then DP23b can not be part of the solution.

Since software and physical components require different kinds of interface access, we need a multi-disciplinary design environment in support of CPSs. In this paper, we exposed the properties of ADT in support of providing techniques which can be utilized in designing CPSs. As a future work, we are planning to extend the ADCO tool [5] with non-functional requirements such as time to measure information content of the entire system.

Acknowledgments I would like to thank UAB Electrical and Computer Engineering Department for providing the visiting scientist opportunity during which I developed this paper. I also thank professor Murat M. Tanik for constructive feedback during development of this work. I also appreciate the invitation from Dr. Urcun J. Tanik to develop to paper.

References

1. E. A. Lee, "Cyber-Physical Systems: Design Challenges," in Object Oriented Real-Time Distributed Computing (ISORC), 2008 11th IEEE International Symposium on, 2008, pp. 363–369.
2. Y. Jia, Z. Zhang, and S. Xie, "Modeling and verification of interactive behavior for cyber-physical systems," in *2011 IEEE 2nd International Conference on Software Engineering and Service Science (ICSESS)*, 2011, pp. 552–555.
3. FlexRay. .
4. C. Togay, A. H. Dogru, and J. U. Tanik, "Systematic Component-Oriented development with Axiomatic Design," *J. Syst. Softw.*, vol. 81, no. 11, pp. 1803–1815, Nov. 2008.
5. C. Togay, "Systematic Component-Oriented Development with Axiomatic Design," Dissertation, Middle East Technical University, Ankara, Turkey, 2008.
6. A. H. Dogru and M. M. Tanik, "A Process Model for Component-Oriented Software Engineering," *IEEE Softw.*, vol. 20, no. 2, pp. 34–41, Mar. 2003.
7. J. Huang, F. B. Bastani, I.-L. Yen, and W. Zhang, "A Framework for Efficient Service Composition in Cyber-Physical Systems," in *Service Oriented System Engineering (SOSE), 2010 Fifth IEEE International Symposium on*, 2010, pp. 291–298.
8. W. Liu, B. Liu, and D. Sun, "A conceptual framework for dynamic manufacturing resource service composition and optimization in service-oriented networked manufacturing," in *2011 International Conference on Cloud and Service Computing (CSC)*, 2011, pp. 118–125.
9. A. Pohl, H. Krumm, F. Holland, I. Luck, and F.-J. Stewing, "Service-Orientation and Flexible Service Binding in Distributed Automation and Control Systems," in *Advanced Information Networking and Applications - Workshops, 2008. AINAW 2008. 22nd International Conference on*, 2008, pp. 1393–1398.
10. J. Liu and L. Zhang, "Aspect-Oriented MDA Development Method for Non-Functional Properties of Cyber Physical Systems," in *Networking and Distributed Computing (ICNDC), 2011 Second International Conference on*, 2011, pp. 149–153.

11. L. Jingyong, Z. Yong, Z. Lichen, and C. Yong, "Applying AOP and MDA to Middleware-Based Distributed Real-Time Embedded Systems Software Process," in *Information Processing, 2009. APCIP 2009. Asia-Pacific Conference on*, 2009, vol. 1, pp. 270–273.

12. "3. Modeling and Analysis of Real-Time and Embedded Systems (MARTE)." [Online]. Available: http://www.omgmarte.org/node/46. [Accessed: 08-Aug-2012].

13. "3. OMG Systems Modeling Language ((OMG) SysML)." [Online]. Available: http://www.omgsysml.org/. [Accessed: 08-Aug-2012].

14. F. Slomka, S. Kollmann, S. Moser, and K. Kempf, "A Multidisciplinary Design Methodology for Cyber-physical Systems," in *Proceedings of the 4th International Workshop on Model Based Architecting and Construction of Embedded Systems*, 2011.

15. P. Derler, E. A. Lee, and A. Sangiovanni-Vincentelli, "Modeling Cyber-Physical Systems," *Proceedings of the IEEE (special issue on CPS)*, vol. 100, no. 1, pp. 13 – 28, Jan. 2012.

16. K. Wan, D. Hughes, K. L. Man, T. Krilavicius, and S. Zou, "Investigation on Composition Mechanisms for Cyber Physical Systems," *International Journal of Design, Analaysis and Tools for Circuits and Systems*, vol. 2, no. 1, p. 30, Aug. 2011.

17. A. Taherkordi, F. Loiret, A. Abdolrazaghi, R. Rouvoy, Q. Le-Trung, and F. Eliassen, "Programming sensor networks using REMORA component model," in *Proceedings of the 6th IEEE international conference on Distributed Computing in Sensor Systems*, Berlin, Heidelberg, 2010, pp. 45–62.

18. "IEEE 1588-2008," 1588-2008 - IEEE Standard for a Precision Clock Synchronization Protocol for Networked Measurement and Control Systems. [Online]. Available: http://standards.ieee.org/findstds/standard/1588-2008.html.

19. Y. Zhao, J. Liu, and E. A. Lee, "A Programming Model for Time-Synchronized Distributed Real-Time Systems," in *Real Time and Embedded Technology and Applications Symposium, 2007. RTAS'07. 13th IEEE*, 2007, pp. 259–268.

20. "IEEE Standard for Modeling and Simulation (M&S) High Level Architecture (HLA)," *IEEE Std 1516 Series*.

21. K. L. Morse, M. Lightner, R. Little, B. Lutz, and R. Scrudder, "Enabling simulation interoperability," *Computer*, vol. 39, no. 1, pp. 115–117, Jan. 2006.

22. "Ptolemy Project." [Online]. Available: http://ptolemy.eecs.berkeley.edu/.

23. N. P. Suh, *Axiomatic Design: Advances and Applications*. Oxford University Press, 2001.

24. C. Togay, O. Aktunc, M. M. Tanik, and A. H. Dogru, "Measurement of Component Congruity for Composition," in *The Ninth World Conference on Integrated Design and Process Technology*, 2006.

25. A. A. Cardenas, S. Amin, and S. Sastry, "Secure Control: Towards Survivable Cyber-Physical Systems," in *Distributed Computing Systems Workshops, 2008. ICDCS'08. 28th International Conference on*, 2008, pp. 495–500.

26. C. Togay, E. S. Caniaz, and A. H. Dogru, "Rule Based Axiomatic Design Theory Guidance for Software Development," in *CORCS 2012: The 4th IEEE International Workshop on - COMPSAC*, 2012.

27. K. Wan, K. L. Man, and D. Hughes, "Specification, Analyzing Challenges and Approaches for Cyber-Physical Systems (CPS)," *Engineering Letters, issue3, EL_18_3_14*, 2010.

28. Y. Zhang, I.-L. Yen, F. B. Bastani, A. T. Tai, and S. Chau, "Optimal Adaptive System Health Monitoring and Diagnosis for Resource Constrained Cyber-Physical Systems," in *Software Reliability Engineering, 2009. ISSRE'09. 20th International Symposium on*, 2009, pp. 51–60.

29. A. Platzer, "Verification of Cyberphysical Transportation Systems," *Intelligent Systems, IEEE*, vol. 24, no. 4, pp. 10–13, Aug. 2009.

30. C. A. Macana, N. Quijano, and E. Mojica-Nava, "A survey on Cyber Physical Energy Systems and their applications on smart grids," in *Innovative Smart Grid Technologies (ISGT Latin America), 2011 IEEE PES Conference on*, 2011, pp. 1–7.

31. J. Sztipanovits, "Composition of Cyber-Physical Systems," in Engineering of Computer-Based Systems, 2007. ECBS'07. 14th Annual IEEE International Conference and Workshops on the, 2007, pp. 3–6.

32. S. Bak, K. Manamcheri, S. Mitra, and M. Caccamo, "Sandboxing Controllers for Cyber-Physical Systems," in *Cyber-Physical Systems (ICCPS), 2011 IEEE/ACM International Conference on*, 2011, pp. 3–12.
33. W. Li, J. Bao, and W. Shen, "Collaborative wireless sensor networks: A survey," in *Systems, Man, and Cybernetics (SMC), 2011 IEEE International Conference on*, 2011, pp. 2614–2619.
34. Q. Shafi, "Cyber-Physical Systems Security: A Brief Survey," in Computational Science and Its Applications (ICCSA), 2012 12th International Conference on, 2012, pp. 146–150.
35. C. Gill, "Cyber-Physical System Software for HCMDSS," in Joint Workshop on High Confidence Medical Devices, Software, and Systems and Medical Device Plug-and-Play Interoperability, 2007. HCMDSS-MDPnP, 2007, pp. 176–177.
36. N. P. Suh, Complexity: Theory And Applications (Mit-Pappalardo Series in Mechanical Engineering). Oxford University Press, 2005.
37. "Acclaro DFSS." [Online]. Available: http://www.axiomaticdesign.com/products/default.asp
38. B. Gumus and A. Ertas, "Requirement Management And Axiomatic Design," *J. Integr. Des. Process Sci.*, vol. 8, no. 4, pp. 19–31, Dec. 2004.
39. D. Tang, R. Zhu, S. Dai, and G. Zhang, "Enhancing axiomatic design with design structure matrix," *Concurrent Engineering Research and Applications*, vol. 17, no. 2, pp. 129–137, 2009.
40. A. Beugnard, J.-M. Jezequel, N. Plouzeau, and D. Watkins, "Making components contract aware," *Computer*, vol. 32, no. 7, pp. 38–45, Jul. 1999.
41. T. Lee, "Complexity Theory in Axiomatic Design," Dissertation, Massachusetts Institute of Technology, Department of Machanical Engineering, 2003.
42. K. C. Kang, S. G. Cohen, J. A. Hess, W. E. Novak, and A. S. Peterson, "Feature-Oriented Domain Analysis (FODA) Feasibility Study," Carnegie-Mellon University Software Engineering Institute, Nov. 1990.

Chapter 9
The Importance of Grain Size in Communication Within Cyber-Physical Systems

Julia M. Taylor

This chapter will look at various applications of natural language communication to cyber-physical systems. One of the assumptions that it makes is that such communication is not only necessary for the future systems, but also should be done on a level acceptable and natural to humans, rather than training them to accommodate machine capabilities with exact and precise commands. We will address a grain size of commands or descriptions that could be given to a system—at the same time the physical capabilities of a system will be sketched only as needed for purposes of examples. The range of commands that we are talking about is a typical algorithmic description of a task at the low level and a natural one for a human task description on the high level. A low, more detailed, fine-grain-sized level is assumed to exist already. The higher, coarser-grain-sized level is what we are striving for, in the sense of being able to switch to it automatically when convenient, i.e., to pay with some vagueness, as people and language do, for the ease of not having to resolve an ambiguity.

One of the more difficult things that are taught in algorithmic-thinking-101 is things that we do every day but don't think about enough to describe them. Outline, for example, a step-by-step process to boil eggs. We pride ourselves on being able to explain it in fine detail and praise a 7-year-old that can describe such a process on their own. There are two questions here that come to mind. The first one is, do we want to communicate with agents or systems on such a detailed level? We will not pretend to answer this question here but rather propose a solution if the answer happens to be "no". The second question is what should happen when a command of *drop the egg in the water* is given. In other words, should the egg be really *dropped*, or should correction for the smooth and slow execution of the command be allowed.

We will start with the latter, easier-to-answer question and work our way to the former, much more difficult question with an outlined solution. We will build on

J. M. Taylor (✉)
Computer and Information Technology, Purdue University, Indiana, USA
e-mail: jtaylor1 @purdue.edu

S. C. Suh et al. (eds.), *Applied Cyber-Physical Systems*,
DOI: 10.1007/978-1-4614-7336-7_9,
© Springer Science+Business Media New York 2014

the methods of the Ontological Semantics Technology, explained in the OST section. We will also briefly explain how such methods of communication could be used in developing-algorithmic-thinking classes.

Perhaps an easier initial example is giving directions. When one person asks another person how to get from A to B, some negotiation of knowledge takes place and information at the appropriate level is delivered. For example, one can say that B is right next to C, assuming that C is known to both people, and that's all that needs to be said, there is no need for turn-by-turn instructions. While this example and the boiling of an egg look somewhat different, they have one important thing in common: what is stated in a "natural" communication between people is actually "instinctively" limited to what the questioner/recipient of information may not know already because what they are all aware of is not necessary to restate.

This need to communicate only the necessary information can be looked at from various angles. One angle is granularity: we can afford to get to the highest/coarsest possible grain size that would activate the finer grain size of information in people's brain, without explicitly stating fine-grain details. Another angle is that of the processing of the unsaid [1, 2], which we will leave the unsaid and its inferences for later explorations and concentrate here on the grain size.

For the purposes of this paper, we will separate what people say into local granularity and global granularity of information. Global granularity will refer to that of a script-like (see [3, 4]—cf. the seminal [5, 6]) phenomena where some of the components of the scripts are well known and not verbalized. In other words, instead of telling a story of several paragraphs, only a couple of sentences are necessary to outline the picture. We will refer to local granularity where what is explicitly stated can be treated as a hypernym or a hyponym of what is actually meant. This distinction is, of course, not black or white and there is a lot of gray area in the middle, for instance where the needed and known information can be supplied in one or two sentences, within which the known details are omitted.

Ontological Semantic Technology

We rely on the Ontological Semantic Technology (OST) [7–13] for the needed grain size interpretation. Ontological Semantics is not the only theory/methodology/technology that can handle what is described here. Any system that has a solid and representative of the world ontology that has enough reasoning capability and that is linked to a lexicon should do the trick. Some part of the system has also to accommodate common sense knowledge that people use in every day communication and a collection of scripts that can be accessed for a given scenario. What knowledge base (ontology or not) contains this information is not necessarily important, as long as it can be accommodated. We use OST for convenience's sake and because it can be easily modified to reflect the needed changes.

Ontological Semantic Technology is one of the next generation systems that the theory of Ontological Semantics [14] has produced. What we describe here is the significantly modified version of previously developed commercial systems. OST is not a domain specific technology: it attempts to work with any topic that a human would typically hold a conversation about. Within that, certain domains have more emphasis, especially if they are in a particular application of interest.

At the core of OST is the ontology (see Fig. 9.1)—a model of the world that encompasses all of the non-instantiated knowledge that is needed to comprehend information exchange between a human and a machine. The ontology is language independent—any natural language communication is interpreted through the concepts and relationships that the ontology contains. Just like any two speakers that are fluent in the same multiple languages can start a conversation in one language, switch it to the next, and the next, and the information that is delivered in any of them is about the same, the ontology has the power of that representation and provides the underlying power of the conversion or comparison. Ontology contains concepts and properties, it outlines relationships between concepts through properties, as well as more complex situations that can be bundled together and be useful at a needed grain size to compress or expand information.

A lexicon is a language dependent inventory of all senses of the words for a particular language. Each sense is described in terms of how it is pronounced, what kind of morphological forms it can take, what part of speech it is, what syntactic constructions it can participate in and, most importantly, what is the meaning of the sense. An onomasticon is a language dependent inventory of proper names that are required to support the application in question for a particular language. Every

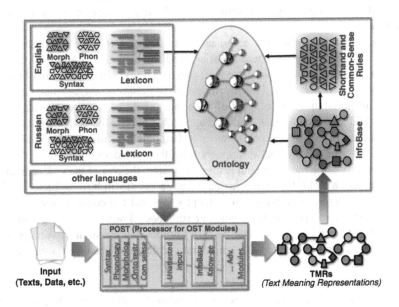

Fig. 9.1 Ontological semantic technology architecture

lexicon and onomasticon sense of a word is anchored in the ontological concept or a "bundle" of concepts that represent the meaning of the sense of the word. What is relevant for cyber-physical systems, is that, through this representation, every command or physical entity loses its ambiguity that is omnipresent in natural language and can be used more precisely. It should be noted that some vagueness will still remain—just like it remains for a human-to-human communication.

Common Sense and Shortcut rules (see [8], cf. other approaches in [14–18]) are rules rather than descriptions and definitions that are in the ontology. Their primary purpose is disambiguation of senses. For example, this repository contains knowledge that you cannot put something larger inside something smaller; that before you end something you have to start it, that a person can be only at one physical location at the same time (grain size is important here too).

Each of these static resources comes with modules that process them. For example, the lexicon comes with syntactic, semantics, phonological and morphological processing. All of these have to return a successful result for the system to return a simple case of a Text Meaning Representation (TMR) of a sentence. It should be noted that for ambiguous sentences there is more than one TMR. A more complicated processing is involved when a word is unknown or when inferences are required. A set of all modules that are responsible for processing information is accessible by the Processor for OST Modules (POST). POST does not only take into account information in text, but may, if needed, look for prior knowledge into InfoBase.

InfoBase is the most "knowledgeable" component of OST—it is where all processed data are stored. InfoBase contains instantiated information of all concepts that were needed to process a particular text. Thus, a generic CAT information is stored in the ontology, but information about particular cats, is stored in InfoBase.

As an example, consider a command "find a kid who knows his name and address." What is of interest here as far as the disambiguation of a lexical sense is concerned is that the word *kid* has at least two meanings, that of a human child and a baby goat (see Fig. 9.2). The restriction of the concept KNOW—that the word *know* is anchored in—for the agent of the event should be an ANIMATE, and both a human child and a baby goat could be applicable here.[1] The restriction of KNOW with a topic of NAME and ADDRESS can only be applicable to a human child [13], thus the sense of a baby goat disappears from the consideration.

Notice that while the command is disambiguated, the rest of the task remains to be performed: a child that knows his name and address still have to be found. We are now looking at the ontological concept KNOW and the common sense knowledge repository of how to check whether somebody knows something and what does it take to find out. From there we will find out that question/answering is the common tool for knowledge solicitation and verification. Therefore, whoever was

---

[1] Let us assume here that animals do know things and can be valid agents of the concept KNOW just like people are valid agents of it.

Fig. 9.2 Kids chasing a kid (http://www.thisistheplace.org/what_we_do/special_events.shtml)

given the command to find a child, should ask a question of every child and listen for the answer. Notice again, that the task is performed in natural language, and further conversation (other than simple question/answer) may be necessary. Also notice that while some answers are received, unless there is known data about the names and the addresses of these children, what is collected by the executor of our command cannot be verified.

Local Granularity: What is it All about and Should We Be Concerned?

Concepts in natural language are typically easier to disambiguate than to find a correct relationship between the words. For example, even though there were 2 types of kids (person and goat) in the picture, the language itself describing the task dictated that the sense of a goat be rejected.

Of course, it is possible that, instead of giving the command of *finding a kid that knows its name and address*, the command of *catching a kid* be given. All of a sudden no ontological restriction can help the disambiguation mechanism: it is possible to catch both children and goats—so what was meant? One way of handling it is to ask a differentiating question: should a child be caught or a goat? Notice, that if the answer to the question is the former, than more questions arise: any child or a particular child? It might be easier to start the negotiation with asking, which kid, but that, again, depends on whether there is an understanding that there is more than one possibility. And the minute we realize that there is more than one possibility, they have to be represented (see Fig. 9.3 for a hierarchy).

For the purposes of this paper, we refer to local granularity as a phenomenon where ambiguity cannot be resolved and it is masked by coarsening the grain size

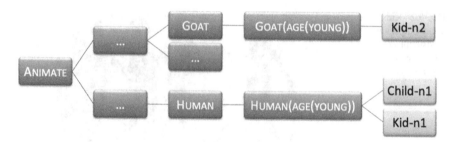

Fig. 9.3 Ontological hierarchy of concepts (in *blue*), with some anchored lexical items (in *green*)

of representation.[2] It is a choice, on the part of the architects, whether to go this route or not. The choice depends not only on the layers of hierarchy between the mutual parents, but also on the number of finer-grain size ontological items in which the ambiguous lexical senses are anchored.

Local granularity phenomenon is even more noticeable in the cases of relationships, especially where words that would be anchored in them are not explicitly stated. A typical example used within OST is that of an *IBM lecture* where it is not clear what the intended relationship between *IBM* and *lecture* is, even if both lexical units used in the phrase are disambiguated. It is tempting to think that information in a sentence would reduce the number of relationships to a minimum, but as demonstrated in [22], it is often not the case. For example, for a sentence *What level of industry expertise exists at local level*, human subjects came up with at least 6 interpretations of *industry experience* within that sentence.

The question of how to represent such vagueness is not just of the semantic nature, especially in the cases of complex nominals (e.g. cat milk bottle). While it is possible to go up the hierarchy of the relationships until the root is reached, it may be just as useful to rely on the syntactic constrains and thus indicate that no semantic restrictions are found at this time [21]. It is possible that this distinction is lost entirely on a native speaker, but it may ease the processing for the machine.

---

[2] The author is grateful to Victor Raskin for pointing out that this is not that different from considerations underlying Weinreich's [19] objections to what he referred to as Katz and Fodor's [20] "infinite polysemy" in their semantic theory. Why, Weinreich asked, does the theory have to differentiate between two senses of *ingest* (eat solids/drink liquids) but not between two senses of *eat* (with a fork/with a spoon)? Reversing it to fit our discussion, we can say that English masks the latter distinction with the word *eat* but reveals the former distinction with two different words, both, incidentally, in much more common usage than the masking *ingest*.

Global Granularity: A More Interesting Case

We start our discussion with an example of brushing teeth: we all do it every day, and we can probably all easily generate an algorithm of how to do it. Yet, if many of these algorithms were collected, most of them would represent information at a different grain size. Why is it that something that we perform so often and in a similar enough manner, produces such a different description? Let us pretend for the purposes of this discussion that we can describe situations and actions without relying on specific types of memories and that these types do not influence the description.

Now suppose, that you run out of toothpaste before you start brushing your teeth. No matter how different the collection of algorithms were, the result is going to be the same: either the teeth will not be brushed, or more toothpaste will have to be found. In other words, all those algorithms, different in details, will halt at the same place. The same could be said if, all of a sudden, one would run out of water before one can rinse their mouth: the procedure would halt at approximately the same place.

Assuming that one can talk, we would probably not explain why we need toothpaste or water, if we were to ask for more. Moreover, we would expect a certain response: a tube (or some other container) of toothpaste, or an amount of water needed to finish, depending on what was asked. We would not need, nor expect, several gallons of water, for example. Thus, with somewhat different algorithms in mind, we can still assist others if a task is familiar, without describing the whole routine.

We would like the machines to follow the same communication scenario: without explaining step by step what is needed to be done, the knowledge of a particular scenario should be accessed and assistance provided with a request naturally understood by a human. It should be enough to state that one ran out of toothpaste, as a command that another tube should be retrieved (similarly to Searle's indirect speech acts in [22]). Notice also, that just like a human should know that if there is no toothpaste in the house, it should be bought, a machine should understand the same thing.

Where will all this knowledge come from? According to OST, generic knowledge of the world rests either in the ontology or in the common sense and shortcut repository. A generic human script for brushing teeth (overlap between descriptions of how to do it) can be entered also in the appropriate location. An individualized script (instantiated) can be adapted to a particular situation and stored in the InfoBase. This instantiated script can be a result of several conversations. When a human signals that (s)he ran out of toothpaste, an ontological concept triggered by the word's appropriate sense would be retrieved from the ontology and a search of the InfoBase would be initiated. Upon retrieving the needed (instantiated and individualized) script from the appropriate repositories, a halting point will be found and a correction would be provided. What is also of interest here is that since OST's ontology is nothing but a graph, all (weighted) links to where toothpaste can be retrieved from will be found, and processed until the solution in the form of toothpaste is found. And thus, if no toothpaste exists at the location that the person is at, a link that connects toothpaste to an event of BUY will be found as well.

Natural Language Communication with Robots: Examples and Analysis

Most situations where a robot would have to react to natural language are likely to be both local and global. We described how to adapt an OST natural language lexicon to a "Robotese" in [23]. We thus assume here that any robot that we work with comes with a lexicon suitable to perform the tasks that it is capable of.

The terms local and global granularity may not be the best terms for the description of the phenomena at hand. Whenever lexical senses create ambiguity that cannot be resolved, but can be represented by a single higher-grain concept, local ambiguity is at play. In other words, the senses are anchored at a fine-grain concept and we climb up to make the description more compact. Global granularity does the opposite: a sense is anchored in a course-grain concept and in order to understand the text, information from finer-grain concepts has to be brought up. It has a top-down flow of representation, rather the climb that happens in the local granularity scenarios.

Consider a scenario where you are communicating with several robots [24], but you wish them to perform commands that they are capable of. Suppose, you have a robot that is on wheels and a humanoid that can jointly perform a task. Also, suppose, that a humanoid can not only walk, but also run. A command: *move to [name your object that they can both recognize or aware of]* should start them on their "journey." Several things of interest here: while you may anchor your function that is responsible for physical movement of a robot in a fairly generic concept MOVE, it is probably a better choice to anchor it in something that corresponds to MOVE (INSTRUMENT (WHEEL)) in the ontology. If that is the case, then a mechanism of global granularity would have to lower the grain size to the needed concept. At the same time, a command that was given only a named object but not a direction of movement or distance, for that matter, has to be adjusted to what the robot expects and needs to receive in order to function.

On the other hand, the humanoid can both run and walk, and it was given a command to move, which is an ancestor of the ontological concepts representing its real capabilities. It can consider several things in order to decide how to move, including its most stable mode of transportation, or the speed of the wheeled robot. Again, the direction would have to be taken into account here, just as it had to be taken into account for the wheeled robot.

Now, suppose the command is *move inside the building and retrieve a table next to the door*. Let us assume that whoever is giving he command is aware of the physical constraints of the robots and will not ask a humanoid to move the piece of furniture—notice a different sense of the word *move* used here—something that it is not capable of. Let us also assume that whatever humanoid can lift can be placed on the wheeled robot and brought it back that way. The question that remains to be answered is what is a table: is it a small enough piece of furniture that is light enough or some kind of a flat surface with a chart on it? While the furniture should have a much higher weight since it is retrieved from the house, and thus has a stronger association with it, the sense of a chart cannot be dismissed either. This means that one command will have both local and global granularity mechanisms in play.

It could be argued that it is unnecessary to account for both local and global granularity and it only overloads the system: it is just as easy to give exact commands to the robots, according to how they are programmed. One could only understand the word *move* as use your wheels to move forward, and for the humanoids to understand a command of movement either *walk* or *run* have to be mentioned. It may also be possible to argue that even if they can tell a furniture table and some flat surface apart from other objects, they could be disambiguated by a human, or count on the fact human could do the work. But then, again, our goal is seamless communication between humans and machines, in a way that is natural to humans.

Our goal is also for a GPS device to be able to negotiate with a person about what they really want to know and voice just that information. It should not only use streets and intersections for such negotiation, but also buildings and other reference points naturally noticeable to a human. And, it should not use a reference point that may appear frequently—not turning after red barn should be mentioned. Finally, it should adjust to what human wants to know at the grain size that is acceptable to a human in communication with another human.

And then, maybe, just maybe, one day it will be good enough to suggest a better solution to a proposed plan.

References

1. J. M Taylor, V. Raskin, C. F. Hempelmann, & S. Attardo, An Unintentional Inference and Ontological Property Defaults. Proc. IEEE SMC 2010, Istanbul, Turkey, 2010.
2. V. Raskin, J. M. Taylor, & C. F. Hempelmann, "Ontological Semantic Technology for Detecting Insider Threat and Social Engineering." *Proc NSPW*-2010, Concorde, MA, 2010.
3. V. Raskin, S. Nirenburg, I. Nirenburg, C. F. Hempelmann, & K. E. Triezenberg, "The Genesis of a Script for Bankruptcy in Ontological Semantics, in G. Hirst and S. Nirenburg, Eds., Proceedings of the Text Meaning Workshop, HLT/NAACL 2003: Human Language Technology and North American Chapter of the Association of Computational Linguistics Conference. ACL: Edmonton, Alberta, Canada, May 31, 2003.
4. J. A. Crowder, J. M. Taylor, & V. Raskin, "Autonomous Creation and Detection of Procedural Memory Scripts." International Conference on Artificial Intelligence, Las Vegas, NE, July 2012.
5. R. C. Schank, Computer Models of Thought and Language. San Francisco: Freeman, 1973.
6. R. C. Schank, & R. Abelson, Scripts, Plans, Goals, and Understanding. New York: Wiley, 1977.

7. V. Raskin, C. F. Hempelmann, & J. M. Taylor, "Guessing vs. Knowing: The Two Approaches to Semantics in Natural Language Processing," in A. E. Kibrik, Ed., Proceedings of Annual International Conference Dialogue, Moscow, Russia: AABBY/Yandex.

8. J. M. Taylor, V. Raskin, & C. F. Hempelmann, "On an Automatic Acquisition Toolbox for Ontologies and Lexicons in Ontological Semantics." Proc. ICAI-2010: International Conference on Artificial Intelligence, Las Vegas, NE, 2010.

9. C. F. Hempelmann, J. M. Taylor, & V. Raskin, "Application-Guided Ontological Engineering." Proc. ICAI-2010: International Conference on Artificial Intelligence, Las Vegas, NE, 2010.

10. J. M. Taylor, V. Raskin, & C. F. Hempelmann, "From Disambiguation Failures to Common-Sense Knowledge Acquisition: A day in the Life of an Ontological Semantic System." Proc. Web Intelligence Conference, Lyon, France, August, 2011.

11. J. M. Taylor, V. Raskin, & C. F. Hempelmann, "Post-Logical Verification of Ontology and Lexicons: The Ontological Semantic Technology Approach." International Conference on Artificial Intelligence, Las Vegas, NE, July, 2011.

12. J. M. Taylor, & V. Raskin, "Graph Decomposition and Its Use for Ontology Verification and Semantic Representation, Intelligent Linguistic Technologies Workshop at International Conference on Artificial Intelligence, Las Vegas, NE, July, 2011.

13. J. M. Taylor, & V. Raskin, "Understanding the unknown: Unattested input processing in natural language," FUZZ-IEEE Conference, Taipei, Taiwan, June, 2011.

14. S. Nirenburg, & V. Raskin, Ontological Semantics. Cambridge, MA: MIT Press, 2004.

15. J. McCarthy, "Programs With Common Sense." Proceedings of the Teddington Conference on the Mechanization of Thought Processes, 75–91. London: Her Majesty's Stationary Office, 1959.

16. D. B. Lenat, "CYC: Toward Programs With Common Sense." *Communications of the ACM* 33:8, 30–49, 1990.

17. J. M. Gordon, & L. K. Schubert, "Quantificational Sharpening of Commonsense Knowledge", in: Havasi 2010, 27–32.

18. C. Havasi, D. B. Lenat, & B. Van Durme, Eds., Commonsense Knowledge: Papers from the AAAI Fall Symposium. Menlo Park, CA: AAAI Press, 2010.

19. H. Liu, & P. Singh, "ConceptNet—a Practical Commonsense Reasoning Tool-Kit, BT Technology Journal, 22:4. 211–226, 2004.

20. U. Weinreich, "Explorations in Semantic Theory." In: T. A. Sebeok, Ed., Current Trends in Linguistics, Vol. 3, 395–477, The Hague: Mouton, 1966.

21. J. J. Katz, & J. A. Fodor, "The Structure of a Semantic Theory," Language 39:1, 170–210, 1963.

22. J. M. Taylor, V. Raskin, & L. M. Stuart, "Matching Human Understanding: Syntax and Semantics Revisited." Proc. IEEE SMC 2012, Seoul, Korea, 2012.

23. E. Matson, J. Taylor, V. Raskin, B.-C. Min, & E. Wilson, "A Natural Language Model for Enabling Human, Agent, Robot and Machine Interaction." The 5th IEEE International Conference on Automation, Robotics and Applications, Wellington, New Zealand, December 2011.

24. D. Erickson, M. DeWees J. Lewis, & E. T. Matson, "Communication for Task Completion with Heterogeneous Robots." 1st International Conference on Robot Intelligence Technology and Applications, Gwangju, South Korea, 2012.

Chapter 10
Focus, Salience, and Priming in Cyber-Physical Intelligence

Victor Raskin

Introduction

Information and language processing in a growing variety of computer applications can achieve the accuracy/precision and reliability that human users require if based on human-like understanding. While statistical and/or machine learning methods in text processing have been perfected in the last two decades (see, for instance, [1, 2] and references there), their successes have been limited in scope, primarily to the clustering tasks, and even there, the applications have excelled better in recall than in precision, leading to customer disappointments with commercial products.

The same last two decades have witnessed the development of the first comprehensive meaning direct access approach in computational semantics. Started in the late 1960s-early 1970s by a small group of computational linguists, free of the fear of semantics, it has remained a minority group still fighting an uphill battle for recognition in the discipline of NLP, always dominated by non- and anti-semantic approaches, syntactic parsing prior to 1990 or so and statistical ever since. Fighting against the mostly uninformed and scared charges of non-feasibility, subjectivity, and high cost of acquiring the "brute force semantic" resources, such as semantic lexicons, Ontological Semantics [3] and its advanced and revised successor, Ontological Semantic Technology [4–10], have demonstrated that various proof-of-concept applications and basic commercial implementations are firmly within reach, given the appropriate informed attitude and preparation. A brief sketch of OST is presented in "Ontological Semantic Technology".

The declared goal of OST is human-like understanding of natural language text. Its practical implementations, still work in progress, have reached the stage of representing reasonably well (and, of course, disambiguating in the process) what the sentence actually means. But humans understand much more than that.

V. Raskin (✉)
Purdue University, 500 Oval Drive, West Lafayette 47906, USA
e-mail: vraskin@purdue.edu

S. C. Suh et al. (eds.), *Applied Cyber-Physical Systems*,
DOI: 10.1007/978-1-4614-7336-7_10,
© Springer Science+Business Media New York 2014

Moreover, they cannot always tell between what a sentence actually means and what they understand. Thus, challenged to explain the meaning of the sentence, *Jack and Jill were married to each other*, an undergraduate tends to tell the whole story of how Jack and Jill:

- first met;
- liked each other;
- went out;
- started dating;
- went exclusive;
- got engaged;
- had a wedding;
- started sharing a household;
- owned property together
- had regular sex;
- might have children; etc.

In fact, however, all of these statements are possible inferences from the original sentence, all quite plausible but every one of them defeasible, and the real meaning of the sentence is that Jack and Jill had undergone some sort of procedure which makes their culture, society, and state recognize them as a married couple. Otherwise, their marriage may have been by proxy, and they had never met in person prior to the ceremony, they may have lived apart, never had sex, etc. OST provides a good foundation for calculating inferences but it has not yet been fully implemented (see also "Ontological Semantic Technology").

Yet another resource that is available to humans for the expansion of a sentential meaning is commonsense knowledge. Thus, the knowledge of the monogamy law adds the inference, which is absolute and indefeasible, that neither Jack nor Jill were married to somebody else at the time they were married unless either committed a crime of bigamy as well as that they are still married now unless they got a divorce or at least one of them died. Recent developments in OST [4–10] shed some additional light on how that knowledge of the world is partially captured in the ontology and partially acquired through "blame assignment," that is the search for reasons for failures in the process of OST analysis, resulting in wrong meaning representations.

INVITE			
	is-a	communicative event	
	agent	human	
	beneficiary	human	
	theme	social-gathering	
	purpose	entertainment	
invite			
	invite	agent	[preceding NP]
		beneficiary	[following NP]
		theme	[to NP]

The third aspect of human understanding is focus: not all elements of meaning are equal—some have more significance for a certain human or group of humans at a certain time. The one isolated sentence about the marital status of Jack and Jill may not be a good example of focus but, even outside of any context, one hearer may be interested in Jack, another in Jill, a third in the incidence of marriage in a certain neighborhood. "Text in Focus" provides a human analysis of a longer text from the point of view of focus.

Text in Focus

Many information and language processing systems are utilized for military purposes. The battleground reports, for instance, may cover all aspects of the situation that a commanding officer or general needs to absorb fast enough to make decisions. Summarizing the most important aspects automatically is useful but risky: the summary must match the commander's focus and provide additional details while the other aspects of the reports may be irrelevant at the moment, and there may be no time for several rounds of additional inquiries and requests for details.

Below is a text from the Internet that we can use instead of a real-life, probably classified report. Every paragraph of the text is followed by its generic focus value in square brackets.

WLAN killing system bound for Iraq
[WLAN weapon for Iraq]
By Lester Haines
[Author: Lester Haines]
The US Army will by June deploy in Iraq its "Matrix" system of remotely-detonated landmines, despite widespread concerns about the technology. The Mosul-based Styker Brigade will, according to Yahoo! news, be able to control individual devices from a laptop via a WLAN set-up. The Army reckons Matrix will eliminate accidental deaths caused by dumb landmines. Critics say otherwise.
[Matrix laptop-controled landmines to deploy in June in Mosul]
Following successful tests in September, the US will deploy 25 sets of mines in Iraq. These include both M18 Claymores, which deliver steel balls, and the "M5 Modular Crowd Control Munition"—a non-lethal rubber-ball-delivering alternative. The Army's Picatinny Arsenal in New Jersey said in a January statement that Matrix was intended for "firebase security, landing zone security, remote offensive attack and both infrastructure and check point protection".
[25 sets of mines to be deployed of two kinds: M18 Claymores delivering steel balls and non-lethal rubber balls. Used both for perimeter security and offense]
Matrix project leader, Major Joe Hitt, declared: "The system is user friendly and a soldier will require a minimal amount of training in order to safely employ and use the system."
[Project leader: minimal traing for operator]
However, Human Rights Watch researcher, Mark Hiznay, countered: "We're concerned the United States is going to field something that has the capability of taking the man out of the loop when engaging the target. Or that we're putting a 19-year-old soldier in the position of pushing a button when a blip shows up on a computer screen."

[Human Rights: Danger of full automation or of a young soldier mistaking a screen blip for enemy personnel]

Globalsecurity.org military analyst, John Pike, weighed in: "If you've got 500 of these mines out there, trying to figure out which one you want to detonate, when the clock's ticking, well that could be a brain teaser."

[Military analyst: Hard to make a quick decision when operating 500 mines]

Several organisations, including Landmine Survivors Network and the Presbytarian Peacemaking Program are urging opponents of Matrix to lobby Donald Rumsfeld for the deployment to be scrapped. A Landmine Survivors Network statement asserts: "It seems obvious that these remote-control anti-personnel mines, however carefully monitored, will present new dangers to innocent Iraqi civilians for years to come."

[Organisations urging to protest deployment as presenting new dangers to civilans]

The US military has made available few technical details about Matrix, or how it works in practice. We at *El Reg* hope that enterprising Iraqi insurgents do not make merry with the Army's Claymore-controlling WLAN. It would certainly give a new edge to the phrase "wardriving".

[Few technical details, warnings of enemy stealing the system]

A human will also conclude that this material focuses on resistance to the development and deployment of more advanced weapon systems, especially, landmines. The document will also provide differently weighted information to those interested in new weapon systems, use of computers in weapon systems, amounts of required military training, possibility of errors in weapon use, etc.

This paper explores the possibility and feasibility of calculating the focused summaries of texts, both generically and specifically for certain biases, and the additional aspects of calculating the focus for cyber-physical devices. The relation of focus to salience and priming will also be discussed.

Ontological Semantic Technology

The precondition for any work on focus or, for that matter, inference or commonsense knowledge, is a fully developed meaning-based computational system capable to represent accurately the meaning of each sentence in a text and, ultimately, of the text as a whole. As we mentioned in previous sections, we use the Ontological Semantic Technology as the theory and methodology for developing such systems.

The first and essential function of OST, and the one that has been implemented most fully, is to interpret a natural language text, sentence by sentence. The OST processor reads each sentence linearly and looks it up, word by word, in the OST English lexicon. Every sense of every (non-auxiliary, non-parametric) word in the lexicon is anchored in an ontological concept, with its properties and fillers, and the fillers can be restricted by the sense. The OST ontology, unlike its lexicons, is language-independent (see, again, [3] for the basic theory of Ontological Semantics, and [4–10] for the much revised OST). This brief sketch borrows some material, including the examples and Fig. 10.1 below, from [11], and this author is grateful to his coauthors for their permission to reuse that material in this section

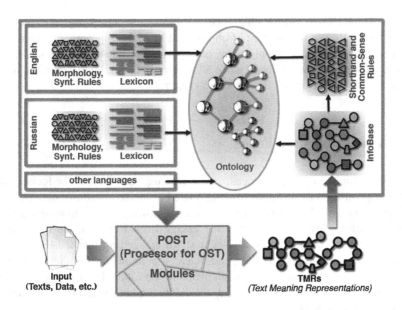

Fig. 10.1 OST architecture

To use a greatly simplified example, the sense of the English word *invite* will be anchored in the ontological concept, probably also labeled "INVITE." The label does not contain any but distinguishing information for the computer and can be any ASCII combination—it is there just for the convenience of the human acquirer.

The text meaning representation (TMR) of the first sentence of the story will result from matching the meaning of the NPs in the appropriate EVENT slots. The reality is, of course, harder, with more complex syntax, ambiguity, etc. The un-enhanced-OST problem with any story is still more advanced: while TMR for each sentence is not hard to produce, the system will not be able to relate the sentences to each other, and the text will lack cohesiveness. With cohesiveness in view, the information processed prior to computing the TMR of the current sentence is used to clarify, complement, and disambiguate the current representation process. This is why we recently added to the OST architecture (see Fig. 10.1) the common sense knowledge resource [9, 12] and the methodology of adding to it when the TMRs fall short of the (often hypothetical) gold standard (cf. [12].

One theory-building question, with methodological consequences, would be whether OST needs a separate resource for common sense knowledge or the rules can be stuffed into ontological concepts. There is a price to pay for either solution, but we have opted for a separate resource in order not to have to duplicate the same information in multiple concepts. Thus, our common sense knowledge includes familiarity with a generic situation, in which, when a hard and heavy object collides with a breakable object (think a rock and a glass window) at a certain speed, the breakage will occur.

The alternative to formulating such a rule in a separate resource and thus making it applicable for any pair of objects that have the appropriate properties is to repeat the rule in the concept for any object on either side of the interaction. But the counterargument would be to make all such objects ontological children of the concept HARD-HEAVY-OBJECT or CAN-BREAK-OBJECT.

Text Focus and User Focus

As we mentioned in "Text in Focus", what was analyzed there is the focus of the text, the generic focus. It is, we believe, related to the novelty/informativeness of the text. Humans do indeed have a sense of novelty and informativeness in communication. Say, Alice receives a message from Bob or reads in the local paper, hard copy or Internet, that Jack and Jill are getting married. How informative is the message for her? It depends, clearly, both on how familiar the topic is and whether the message adds new elements to the topic. The range of possibilities, from totally uninformative and ascending to highly informative, can be roughly represented as follows:

- Totally uninformative: Alice already knows that Jack and Jill are getting married (no new information is supplied);
- Uninformative: Alice does not know Jack and Jill and is uninterested in local or global matrimony;
- Very weakly informative: Alice is somewhat interested in upcoming weddings anywhere;
- Weakly informative: Alice is interested in local weddings
- Less weakly informative: Alice is interested in upcoming local weddings and knows (about) Jack or Jill;
- Informative: Alice is interested in upcoming local weddings and knows (about) both Jack or Jill;
- Highly informative: Alice is interested in upcoming local weddings, knows both Jack or Jill as well as about their relationship and has been wondering if they are planning to get married any time soon;
- Maximally informative: Not only is Alice interested in upcoming local weddings, knows both Jack or Jill as well as about their relationship and has been wondering if they are planning to get married any time soon but she also must attend the wedding and buy a present.
- Catastrophically informative: Alice has been romantically involved with Jack and hoping to marry him herself.

In the last but one case, any text about the place and date/time of the wedding as well registry information will keep its maximal informativeness for Alice. In the last case, any detail about the wedding will be of the most acute interest to Alice, who may actually try and sabotage it, one way or another.

Focus is, however, distinct from informativeness. Even a totally uninformative text will have a focus of sorts—at least in the degenerate sense that it will be about something and it will say a few things about that object or event. Arguably, there may be totally unfocused text—simply poorly composed—that talk about several things, none of them obviously connected to each other.

Text focus is generic in a sense that, for any reader, some of that will remain in place whenever the text is processed by him or her. Every individual reader, however, may add to that a specific interest that may supplement or even completely replace the generic focus. Thus, if one is interested only in the extent this paper uses linguistic semantics, one will focus mostly on the section that briefly introduces OST and, possibly, on every mention of OST throughout the paper. If one is interested in the occult, nothing in this paper will correspond to such a user focus. If yet again, one wants to check out the latest work on, say, salience, one will focus on post-2005 references to this and related notions in the References, double-checking on where and how those references are cited in the paper.

That Catch-All Context Disclaimer

There is a clear give-away in a paper or a proposal claiming to be semantic: the use of the word 'context,' as in "it depends on the context": either you know what that context is and how exactly it determines the meaning of your sentence or text or you are not doing real semantics (see an angry statement on the distinction between "real semantics" and the rest in Hempelmann and Raskin 2008).

This fallacy shares the pedigree with the entire half-century history of sentential semantics. Founding it as "transformational/generative semantics," the semantic component of what soon became known as Chomsky's "standard theory" [13], short-lived but with a huge, mostly negative legacy of 'Linguistic Wars" (see [14] for a comprehensive but much-questioned account) Katz and Fodor's [15] semantic theory resided on two directly contradictory statements. The first one, truly innovative and essentially correct, was that the main function of a semantic theory is to model the native speaker's semantic competence, primarily in matching his or her ability to grasp each of the alternative meanings of a sentence, along with catching semantic anomaly, if any, as well as the paraphrasing ability.

The second one came as an under-the-breath disclaimer: the theory was unable to capture the meaning of each sentence in each context. The result was that while the native speaker knows the difference among *taking Junior/the lion/the bus back to the Zoo* by matching the 3 versions with the 3 most appropriate interpretations: Junior wants to go there again for amusement; the lion needs to be returned to its cage after having been rented out for a movie or a bachelor party; the bus will deliver the individual or group there as public transportation, the theory does not.

Rather than trying to capture these differences, Katz and Fodor declare that, to provide and factor in the contextual information they must provide a complete pre-structured description of everything in the world, which they glibly dismiss as undoable. Accordingly, they treat each of the phrases (or sentences with them) as 3-way ambiguous each, and that is not what the native speaker's intuition about them is. Even more strongly, neither of the sentences *Our store sells alligator/ horse shoes* is ambiguous either.

Strangely, nobody but this author ever exposed this essential fallacy that practically nullifies both the descriptive and especially explanatory power of their theory. His Ontological Semantics of the 1990s resulted from decades of research aiming at overcoming the fallacy and thus rendering "context-free" semantics into as contextual as it is for the native speaker. The OST ontology is a structured, formal theory of the world. It is an engineering ontology that is tested, evaluated, and justified in applications rather than making philosophical claims that the world is as it shows or psychophysiological claims that these ontologies are in the minds of native speakers, hardwired and/or induced in their brains, though none of these possibilities has to be excluded (cf. [3]: "Radically Simplifying Cyber Security" and referenced there).

One of the capabilities and functions of the ontology is to provide the initial, basic clues for activating a context. It does this with its properties that directly connect, each of them, two or more nodes. Thus, going back to the ontological concept INVITE, any word (of any language) that is anchored in this concept will automatically evoke communication among humans and a social occasion which is not open to the general public. It may be argued that entertainment, as per that concept, is not the only purpose of a social gathering but just about any other purpose can be achieved without such a gathering. Each of the evoked concepts may, in turn, evoke other concepts but perhaps it can be postulated that the strength of the association will weaken with every extra link.

Besides, memories and experiences may be evoked as well, often whole situations, or scripts, or scenarios. These are a separate interesting and difficult category, which will be left out from this paper (see, however, [16]). People may think of successful communications and misunderstandings, being happy to be invited to parties they wanted to go and dreading invitations to boring or unpleasant gatherings, accepting and declining invitations, working on mental or paper lists, the embarrassment of forgetting to issue or to honor an invitation, the horror of leaving one at home and being suspected of gate-crashing or worse, of being refused admission. All of these experiences and memories would be of invitations as instances of what has happened or may happen. Extracted from those, there are pieces of general ontological truth which may not have been accommodated in the existing ontology. Some properties may be easily enough added or adjusted but others are harder to accommodate. What, for instance, should we add and to what concepts to reflect the fact that we invite people we would like to see at a gathering as well as people we must invite because, for instance, we "owe" them. This piece of knowledge is a good candidate for commonsense knowledge that was introduced as a separate resource in the brief outline of OST.

Context and Focus

So ontological properties provide the initial clues for the general, possibly even generic context that words anchored in a concept evoke. That context is large, and in principle, because the ontology is a connected graph, will include everything in the world; practically, most of that everything will be too remote to determine the meaning. But focus is a much smaller part of that general context, highlighted as it were at any time and also shifting, at least partly, as the text develops. Thus, focus has the more permanent, stable, universal part and the more dynamic, contingent part. The former part contains topics of survival and necessity, such as life and death, danger and safety, keeping and improving one's face (as in "not losing face"—see [17]), stature, and material well-being, fulfilling one's obligations and societal expectations, gaining and keeping the sexual condition that one needs and/ or wants. To give a primitive but illustrative example, in the middle of any involved and involving discussion or argument, seemingly requiring one's full attention, "Fire!" screamed by anybody outside will immediately trigger the flight; in other words, reverting the highly dynamic focus to any part of the permanent focus, no matter how latent at the moment, is pretty immediate and effortless.

The permanent part of focus cannot be ignored because doing so will leave some clear phenomena, like that fire alarm, unexplained. It is, however, the dynamic part that we are focusing on (pun intended) in this paper, and the reason for that is that it determines the meaning in any communication while its content is shifting as the communication progresses. It becomes a major factor in interpreting the text correctly, including hints, insinuations, leading questions, coercion, and propaganda. Thus, a single woman, Alice, haranguing a single man, Bob, that he cannot be as smug as he is about his achievements while he has no wife nor children is, in all likelihood, after Bob as a potential husband for herself and her choice of the father for her future children. Additional knowledge about Alice that she has just come out of a loveless marriage with no issue (by choice) and that she is approaching the age of lower fertility and lower chance of meeting somebody suitable, will confirm this focus, even if never definitively.

This knowledge is not easily accessible to all, and the neutral observer, which is what most people are, those uninterested in either the haranguing woman nor the harangued man, just don't care enough to go so deeply. But there are situations when it is necessary to do it in general and not practical to do it other than computationally, e.g., in detecting and exposing an artful inside traitor [18], who is often an independently frustrated employee, whose unhappiness may have rendered her malicious. So what are the focusing clues that can be programmed into the computer. What are the low-hanging fruit should be the first question. Where do we go beyond the ontological properties?

This paper has a title that announces what it is about. None of the entities there, focus, salience, priming, things cyber-physical, or intelligence is self-obvious. So it stands to reason that this text will address these issues and that every sentence and paragraph should be somehow related to at least one of them. Yet, so far, only one of these entities, focus, has been explicitly mentioned. Moreover, some paragraphs are more remote from focus because they support some other paragraphs that are more related. The focus on focus, so far, has been informing the reader that everything is related to it and that, if the relation is not obvious, more thinking, on their part, is required. The absence of an explicit reference to the other entities in the title indicates to the reader that those will be introduced before the paper is over. The undesirable alternative, if they are not mentioned, is to get a message from the author, intended or not, that somehow the readers should be able to discover the connections on their own. Chances are they would refuse to do so and simply declare the paper or at least its title inadequate.

In a dialog, the focus is prompted by questions that may require more in-depth information or shift the interlocutor to the topic that interests the questioner. Thus, in the example of the harangue, Alice may try and switch Bob from general facts to herself or himself as a topic and evoke emotions. Similarly, in search systems, the users will type in keywords or even real questions (that the system will secretly or openly convert to keywords), correcting, narrowing down, broadening or redirecting their search as they read and evaluate the responses to the previous queries.

In a real-life conversation, such as between Alice, interested in Bob, who does not reciprocate the interest, she will show an interest in any woman Bob happens to mention. She may venture her own opinions of the women they both know in order to solicit his. Each such name changes and adds to the focus. Bob, if he is paying attention, may figure out the reason for this shifting focus and, thus, the underlying more general focus. He can then either challenge Alice or finish the conversation or, more subtly, change the subject. In turn, Alice, if she is just as smart, will figure out that her underlying focus is exposed and exercise one of the same options that Bob has, herself.

One can see these as "games people play" but they are essential for understanding human communication (see, for instance, [19, 20]). Can it be done computationally? Can inference and reasoning be done computationally? The community of scholars vote yes by continuing to improve their techniques, and OST affords many more opportunities for that because it provides access to meaning. Inference and reasoning are done by humans on this basis, and artificial intelligence, at least in one of its guises, attempts to emulate what people do, even if not necessarily how they do it. We should note also that the shifting focus in Alice's contribution to the conversation is dominated by a permanent focus of sexual/marital achievement, and that is also valuable information for calculating the shifting focus of the moment.

Focus and Salience

This discussion would not be complete without relating focus and context to salience. The last term is used, misused, and abused widely in psychology and neurophysiology but also in communication and semiotics—it has hardly any currency in and around linguistics (see [21, 22]). Perhaps the most contentful interpretation of the term is that what is prominent in the mind at the moment is salient. It is often confused with 'important,' which is acceptable only in the sense of 'important at the moment.'

In neuroscience, a salient property is what makes an entity that possesses it stand out of the lot of surrounding or similar entities (see, for instance, [23] and references there). Both in communication and semiotics (see, for instance, [24–26]), it is used to indicate relevance and/or prominence; in the latter field, it usually characterizes the most significant part or aspect of a sign.

The relation to focus should be obvious: what is prominent and relevant in my mind should affect my focus—this should be what I am interested in, what I seek information on, what the purpose of my activity and communication is, and what, ultimately, affects my comprehension. In reality, it is not necessarily so: I may be worried about my latest blood test while actively continuing my explorations in computational semantics, and the one will have no affect on my comprehension of the other, even though it may lead to an ambiguity: if an associate accepts a phone call and then exclaims, "it is very high," I may misinterpret the phone call as coming from an assistant who just calculated a metric on our latest experiment or as coming from my physician's office and reporting on some alarmingly high count in my blood.

What is means is a further narrowing of the scope in calculating the focus:

- First, there is the permanent focus areas, sometimes designated broadly as "self-interest," or "vested interest" [25].
- Second, the general context is brought up by the ontology and it narrows the permanent focus area(s) to that or those compatible with the is context: the occurrence of a word brings up the concept that the appropriate sense is anchored in; then the concept brings up its properties and other concepts as their fillers—and it may spread into properties of those concepts, their fillers, etc., weakening significantly.
- Third, additional words bring up similar context information, narrowing the focus to constantly recalculated overlaps, or intersections.
- Fourth, related memories and experiences are brought up from the common-sense knowledge resource.
- Fifth, especially in multitasking among various activities or mini-multitasking, as it were, dealing with aspects of the same activities, focus can be further refined grain-size-wise [27]—besides or instead of further restriction, focus can also be shifted to something different though usually somewhat related.

Paired with the notion of salience, the triggering of prominence and relevance is referred to as priming, so starting with the second stage of focus creation, priming keeps occurring. Parallel to other zero-phenomena, zero-priming may occur in text and in situations, and it may have important consequences—as we demonstrated earlier [18, 22], it may have the effect of reconsidering various communication defaults and lead to the detection of unintended inferences.

It should be made clear that the direct relationship between focus and salience does not necessarily mean that the research into either phenomenon can gain from considering them together: that salience underlies focus and focus depends on salience has not really helped us to understand either—pretty much the same questions have remained about either, and pooling the resources of psychology for salience and linguistics for focus has not yielded any additional insights, but it is good to be receptive to a possible interaction.

Focus in Cyber-Physical Systems

Finally, a word on what focus looks like in cyber-physical systems. The only difference they make is alleviating the problem by restricting focus further by their physical limitations that may reduce their cognitive capacities. Assuming incorrectly, for the sake of convenience, that human cognitive capacities are unlimited, any artificial cyber-physical system is limited in comparison.

Raskin [28] examined the world of a robot and indicated that it—and the ontology that reflects it—is not so simple even when a robot's cognitive capacity is limited to one or any small number of sensors, such as measuring proximity and, say, temperature. To that, all of the robot's physical states, notion rules, balance, etc., should be added, plus some other, often unexpected elements of the general ontology reflecting much richer worlds.

Nevertheless, the sensors and their readings will never be remote from the cyber-physical systems' focus, thus significantly restricting the initial permanent, "self-interest" focus. So while one should expect pretty much the same mechanism of shifting-focus calculation, all the values should be expected to be much lower in scope. Much focus can, therefore, be seen as built into a cyber-physical system. Trivially, we can declare the focus mechanism to be built into humans as well: granted, of course, but the difference is that, to a very large extent, we build the cyber-physical systems and thus can claim a much greater, if not full understanding of its design, the "2001" fears notwithstanding.

Conclusion

In this chapter, we have examined the complex and important phenomenon of focus in communication and, primarily, in natural language text. We have also indicated ways to detect focus in computational systems based on direct access to comprehensive meaning. We are leaving the door to statistical and machine learning based approaches as well but not initially. They should work best on a semantically enriched knowledge base, and on this point, we find ourselves in surprising agreement with Minsky [29].

References

1. C. D. Manning, and H. Schütze, Foundations of Statistical Natural Language Processing. Cambridge, MA: MIT Press, 1999.
2. C. D. Manning, P. Raghavan, & H. Schütze, An Introduction to Information Retrieval. Cambridge, UK: Cambridge University Press, 2008.
3. S. Nirenburg, & V. Raskin, Ontological Semantics. Cambridge, MA: MIT Press, 2004.
4. V. Raskin, C. F. Hempelmann, & J. M. Taylor, "Guessing vs. Knowing: The Two Approaches to Semantics in Natural Language Processing," in A. E. Kibrik, Ed., Proceedings of Annual International Conference Dialogue, Moscow, Russia: AABBY/Yandex.
5. J. M. Taylor, V. Raskin, & C. F. Hempelmann, "On an Automatic Acquisition Toolbox for Ontologies and Lexicons in Ontological Semantics." Proc. ICAI-2010: International Conference on Artificial Intelligence, Las Vegas, NE, 2010.
6. C. F. Hempelmann, J. M. Taylor, & V. Raskin, "Application-Guided Ontological Engineering," Proc. ICAI-2010: International Conference on Artificial Intelligence, Las Vegas, NE, 2010.
7. J. M. Taylor, V. Raskin, & C. F. Hempelmann, "From Disambiguation Failures to Common-Sense Knowledge Acquisition: A day in the Life of an Ontological Semantic System," Proc. Web Iintelligence Conference, Lyon, France, August, 2011.
8. J. M. Taylor, V. Raskin, & C. F. Hempelmann, "Post-Logical Verification of Ontology and Lexicons: The Ontological Semantic Technology Approach," International Conference on Artificial Intelligence, Las Vegas, NE, July, 2011.
9. J. M. Taylor, & V. Raskin, "Graph Decomposition and Its Use for Ontology Verification and Semantic Representation," Intelligent Linguistic Technologies Workshop at International Conference on Artificial Intelligence, Las Vegas, NE, July, 2011.
10. J. M. Taylor, & V. Raskin, "Understanding the Unknown: Unattested Input Processing in Natural Language," FUZZ-IEEE Conference, Taipei, Taiwan, June, 2011.
11. J. M. Taylor, V. Raskin, & L. M. Stuart, "Matching Human Understanding: Syntax and Semantics Revisited." Proc. IEEE SMC 2012, Seoul, Korea, 2012.
12. J. F. Allen, M. Swift, & W. de Beaumont, "Deep Semantic Analysis of Text," Symposium on Semantics in Systems for Text Processing (STEP). Venice, Italy, 2008.
13. N. Chomsky, Aspects of the Theory of Syntax, Cambridge, MA: MIT Press, 1965.
14. R. A. Harris, The Linguistic Wars. London-New York: Oxford University Press, 1995.
15. J. J. Katz, & J. A. Fodor, "The Structure of a Semantic Theory," Language 39:1, 170–210, 1963.
16. V. Raskin, S. Nirenburg, I. Nirenburg, C. F. Hempelmann, & K. E. Triezenberg, "The Genesis of a Script for Bankruptcy in Ontological Semantics," in G. Hirst and S. Nirenburg, Eds., Proceedings of the Text Meaning Workshop, HLT/NAACL 2003: Human Language Technology and North American Chapter of the Association of Computational Linguistics Conference. ACL: Edmonton, Alberta, Canada, May 31, 2003.
17. P. Brown and S. Levinson, Politeness: Some Universals in Language Usage. Cambridge, UK: Cambridge University Press, 1987.
18. V. Raskin, J. M. Taylor, & C. F. Hempelmann, "Ontological Semantic Technology for Detecting Insider Threat and Social Engineering," Proc NSPW-2010, Concorde, MA, 2010.
19. E. Berne, Games People Play – The Basic Hand Book of Transactional Analysis. New York: Ballantine Books, 1964.
20. E. Goffman, Frame Analysis, New York: Harper & Row, 1974.
21. S. Attardo, "The Role of Affordances at the Semantics/Pragmatics Boundary," in B. G. Bara, L. Barsalou and M. Bucciarelli, Eds., Proceedings of the CogSci 2005: XXVII Annual Conference of the Cognitive Science Society. Mahwah, NJ: Lawrence Erlbaum, 169–174, 2005.
22. J. M Taylor, V. Raskin, C. F. Hempelmann, & S. Attardo, "An Unintentional Inference and Ontological Property Defaults," Proc. IEEE SMC 2010, Istanbul, Turkey, 2010.

23. J. Li, M. D. Levine, X. An, X. Xu, & H. He, "Visual Saliency Based on Scale-Space Analysis in the Frequency Domain". *IEEE Trans Pattern Anal Mach Intell.*
24. G. Bedny, & W. Karwowski, "Meaning and Sense in Activity Theory and Their Role in the Study of Human Performance". *International Journal of Ergonomics and Human Factors* 26:2, 121–140, 2004.
25. W. D. Crano, "Attitude Strength and Vested Interest." In R. E. Petty & J. A. Krosnick (Eds.), Attitude strength: Antecedents and Consequences. Mahwah, NJ: Erlbaum, 131–158, 1995.
26. M. Humphreys, & J. Garry, "Thinking About Salience." Early drafts from Columbia, 1–55, 2000.
27. J. M. Taylor, "The Importance of Grain Size in Communication Within Cyber-Physical Systems," this volume.
28. V. Raskin, "Not So Simple Ontology of a Primitive Robot," RITA-2012: The First Annual Conference on Robot Intelligence Technology and Applications, Gwangju, Korea, December, 2012.
29. M. Minsky, Emotion Machine: Commonsense Thinking, Artificial Intelligence, and the Future of the Human Mind. Presentation at MIT World, 2007, http://mitworld.mit.edu/video/484.

Chapter 11
An Adaptive Cyber-Physical System Framework for Cyber-Physical Systems Design Automation

U. John Tanik and Angelyn Begley

Introduction

This chapter on the dissertation of U. John Tanik [1] by describing an automated approach to Cyber-Physical Systems design utilizing an Adaptive Cyber-Physical System Framework (ACPSF). The ACPSF is based on the Artificial Intelligence Design Framework (AIDF) supported by a NASA training grant from 2004 to 2006 at UAB [2]. The case study for the AIDF shows how a KBE can be configured to provide automated design support for a Free Space Optical Interconnect (FSOI), an example of a Cyber-Physical System. In 2008, VDM Verlag published the dissertation as a book *Architecting Automated Design Systems* [3] demonstrating broad impact by describing how the AIDF can provide effective guidelines for design automation. The ACPSF builds on the experience gained from the AIDF describing how it can be expanded to support Cyber-Physical System design automation. Similar to its predecessor architectural framework (AIDF), the ACPSF provides a comprehensive, modular, reconfigurable, and scalable approach to develop networked knowledge-based engineering (KBE) systems that support the automated design of cyber-physical systems of all types. The same case study is used to show how optical backplane engineering can serve as an example of a CPS artifact designed based on ACPSF guidelines to generate the requirement specifications for a KBE system to support the field of optical interconnects. The adapted KBE system for FSOI provides optimized recommendations based on artificial intelligence, design theory, case based reasoning methods, and other expert knowledge advanced, updated, and stored in the web-enabled knowledge base.

Advances in Cyber-Physical System (CPS) design are emerging as a technology disrupter profoundly impacting the economies of nations [4]. No longer will it be

U. J. Tanik (✉) · A. Begley
Department of Computer Science, Indiana University–Purdue University Indianapolis,
Coliseum Blvd E, Fort Wayne, IN 46805-1499, USA
e-mail: jtanik@gmail.com

S. C. Suh et al. (eds.), *Applied Cyber-Physical Systems*,
DOI: 10.1007/978-1-4614-7336-7_11,
© Springer Science+Business Media New York 2014

adequate to utilize the industrial trends that defined the 20th century to compete with global competitors, rather a much more universally integrated approach must be taken to maintain sustainability. These new challenges require a new integrated approach to multi-disciplinary system design trends that take into account factors impacting both the cyber and physical world simultaneously [5].

While Cyber-Physical Systems are being considered heavily in the research realm, there is a noticeable lack in addressing the numerous design difficulties that present themselves in the Cyber-Physical System design environment. For instance, a unique property of a Cyber-Physical System is the close integration of all aspects contained within the Cyber-Physical System [5–8]. A critical codependence is thus presented to the system design process at all layers and phases, presenting unprecedented and challenging design concerns. The design, construction, verification, and validation of cyber-physical systems pose a multitude of technical challenges that must be addressed by a cross-disciplinary community of researchers, engineers, scientists, and educators [6]. Facing a multi-disciplinary problem must be given a fresh perspective on the design process and Science, Technology, Engineering, and Math (STEM) careers in CPS need to be promoted to meet the global economic needs.

Interactions among the dispersed disciplines generate increasing system functionality, which requires more intense focus with regard to the validation and verification process. The validation and verification process must be both comprehensive and adaptable to evolving environmental conditions, fulfilling both end-user needs and the engineering specifications with special emphasis on dependencies and processing speed. Further compounding the issue of increasing and interdependent functional and nonfunctional quality requirements is the strict restrictions imposed by CPS concerns, such as timing concerns for component reliability [5].

There is a recognized need for an overarching architectural framework to be established to achieve intellectual control over a Knowledge-based Engineering System assisting in the design process for CPS product engineering. This type of approach provides a platform to structure the large-scale software engineering development process for assuring risk mitigation during automated reliability engineering, with applications to National Aeronautics and Space Administration (NASA) [9]. The AIDF formed the basis in 2006 to design CPS artifacts similar to optical backplanes, so a case study is provided relating how it can be supported by the ACPSF. The expanded framework for CPS design has been called the Adaptive Cyber-Physical System Framework (ACPSF) in this paper. Our proposed framework coupled with the Cyber-Physical System methodologies will enable great versatility during the design process by leveraging the global expertise in embedded system design via knowledge engineering methodologies. Topics beyond optical backplane engineering include addressing needs in Aerospace, Agriculture, Energy, Healthcare, and numerous other fields.

Background

This chapter present day technology areas discussed within this section. There are essentially three main concepts contained in the background of this research: knowledge-based engineering, artificial intelligence, and artificial intelligence design framework (AIDF). These three areas together form the basis to the research study on improving design automation for CPS artifacts.

Knowledge-Based Engineering

Knowledge-Based Engineering plays a key role in managing the vital resource of knowledge in a means that is easily accessible to decision making [10]. Such Knowledge-Based Engineering Systems are applicable to different ranks within an organization, ranging from the lower-level call center help desk environment to the decision-making engineering and executive level in the company. A centralized set of knowledge repositories are stored within a database to supply data access to the Knowledge-Based Engineering Management System [11]. While performing daily activities any respective worker can access the data contained within these knowledge repositories via the Knowledge-Based Engineering Management System.

Updates to these knowledge-based repositories are conducted periodically by experts in the respective field. Experts simply enter the new updates through a GUI supplied to them as part of the Knowledge-Based Engineering Management System. These updates are then typically instantaneously propagated through the Knowledge-Based Engineering Management System down to the workers that put the new data into action throughout their daily activities. The end workers are also presented with their own version of a GUI to access the information contained within the KBEM. Due to this efficient sharing of knowledge, businesses are able to relatively quickly adapt to ever-increasing needs presented from the consumer base or engineers are able to procure valuable design knowledge from the factory floor or expert knowledge repositories, which are continuously updated.

Knowledge-based systems can be implemented inside a SQL database system to provide assistance to these workers in the field based upon data provided directly from experts in the respective field. The source data therefore is limited to being input directly from an expert as opposed to considering real-time environmental factors. Modifications can be made to the concepts contained within the common day Knowledge-based system to make it pertinent to other applications [10]. In order to fully utilize knowledge, it has been recognized these Knowledge-Based Engineering Management Systems must be greatly broadened beyond a simple organizational-wide approach [10, 12]. The current global economy pushes businesses to incorporate even the smallest-scale competitive advantage to achieve profitability. Business productivity adjustments made in one industry may also assist another industry in becoming much more efficient. Accessing information of this nature would greatly aid companies with efficiency in their current field, which could be accomplished with an expanded role for Knowledge-Based Engineering Management Systems.

Artificial Intelligence

Artificial Intelligence is applying logical decision making steps within computer-based systems in an effort to allow the computer to make intelligent decisions semi-akin to those a human being makes every day. Logic in this form can today be found in many key devices ranging from Inertial Navigation Systems (INS) [13] to speech recognition systems [14]. On a broad-based scale there are several research ideas for utilizing Web Ontology Languages (OWL) to extend the benefits on current systems [15]. The OWL makes it possible to conduct Artificial Intelligence in various applications based on an Internet standard leveraging Semantic Web advances, thereby yielding a much easier to comprehend system society-wide. Logic for Artificial Intelligence traditionally carries its roots back to first order logic. First order logic is a formal system utilized in both the Computer Science and Mathematics disciplines, respectively. Quantified variables are the core-focus of first order logic, which permits consideration of variables bound by a certain maximum and/or minimum numeric value. Predicates within first order logic are typically contained within sets [16]. One particular example utilizing the first-order-logic can be seen in Axiomatic Systems. Essentially, an axiomatic system can be defined as a set of axioms from which a subset or at times the entire set of axioms can be combined in an effort to logically arrive at theorems, forming design theory [17]. Given the topic focus is on design aspects with Cyber-Physical Systems, we have elected to pair with an Axiomatic Design System tool such as Acclaro Design for Six Sigma [18].

Artificial Intelligence Design Framework

The requirements elications process for the AIDF was managed by the Acclaro Design for Six Sigma (DFSS) tool by Axiomatic Design Solutions. Acclaro DFSS assists in the requirements elicitation and management process for risk mitigation and design parameter identification. Features utilized in the tool to construct the architectural framework for the AIDF include the design matrix and dependency structure matrix functionality, in addition to the Quality Function Deployment (QFD) and the Failure Mode Effects Analysis (FMEA) functionality especially useful for nonfunctional requirements as well. Acclaro DFSS software implements a complete suite of DFSS tools using an axiomatic framework to reduce development risk, cost and time, by applying the axiomatic design process developed at MIT [18].

A core infrastructure has been proposed and conceptualized to handle design of complex systems based on an architectural framework. The framework has been coined as Artificial Intelligence Design Framework (AIDF) and published as a book called Architecting Automated Design Systems [2]. AIDF has been designed in a manner to allow the incorporation of inputs from a real time environment connected to Web Services and networked knowledge bases to be fed into the reconfigured KBE system according CPS artifact design domain. This is then utilized to choose appropriate actions for adjusting the overall design of the system at key trade-off decision points utilizing synthesis and analysis techniques that invoke any of 20 modules that contain knowledge in the form of design theory and related reasoning algorithms acquired from global expert resources. Given the continuous feedback from updated knowledge repositories, the KBE system is able to automatically provide designers immediate feedback on best CPS design practices.

The AIDF and now the ACPSF provide an overall architectural framework and guidelines to develop a respective KBE system for design automation, as in the case for optical backplane design. One aspect relies on knowledge gained from field experts, which is gathered and stored into a knowledge-base. An interface is provided within the system to access this input from the knowledge-base and supply the appropriate data to the artificial intelligence interface for appropriate processing. A supplemental input interface is present in the AIDF as well to permit workers within the environment to provide feedback via Web Services based on the current best practices within that particular facility.

Cyber-Physical Systems Applications

In the simplest sense, Cyber-Physical Systems (CPS), are integrations of computation, networking, and physical processes [8]. An illustration on the flow among a Cyber-Physical System can be seen in Fig. 11.1, with networking providing the communication end as depicted. There are three key characteristics to a CPS with respect to computation/information processing, and physical processes: functionality and salient system characteristics; and computers, networks, devices and their environments in which they are embedded. These core characteristics have been summarized in Table 11.1.

Cyber-Physical Systems contain embedded systems with respect to the computation side involved within the system. Embedded systems have become rather prevalent within today's society; however we have merely scratched the surface of their true utilization potential. Standalone embedded systems are radically restricted in terms of system goals and activities. An overlap occurs among embedded systems and Cyber-Physical Systems, as well as real-time systems as illustrated in Fig. 11.2 [6]. Paired with more robust tools, such as a Knowledge-based Engineering System integrating Artificial Intelligence, embedded systems are transformed into a much more potent type of system, a Cyber-Physical System.

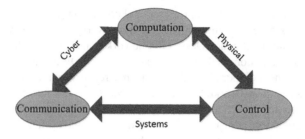

Fig. 11.1 CPS integration overview [4, 27]

Table 11.1 Core CPS characteristics [4]

CPS features	
Computation/information processing, and physical processes	Tightly integrated to the point it is not possible to identify whether behavioral attributes are the result of computations (computer programs), physical laws, or both working together
Functionality and salient system characteristics	Emerging through the interaction of physical and computational objects
Computers, networks, devices and their environments in which they are embedded	Interacting physical properties exist that consume resources and contribute to the overall system behavior

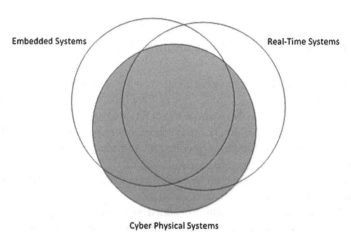

Fig. 11.2 Overlap among cyber-physical systems, embedded systems, and real-time systems [6]

Although there is vast potential for the Cyber-Physical Systems in the future, there are many challenges to overcome in the design of CPS that the ACPSF is expected to mitigate, simplify, and improve. Today's application-oriented organization of the design automation industry has led to expensive, monolithic tool chains that support only a narrow range of problems. These stove piped tool chains do not scale to the large, heterogeneous systems that we need to build using CPS techniques [4]. Designing systems directly geared to a singular application results in a significantly limited system that is not able to be adapted to meet needs of any other industry or oftentimes even an additional business within the same industry. Hence an ACPSF can provide a framework to adapt KBE systems to appropriate support for any CPS domain when the related domain modules are loaded. There are two key points to component based systems resulting in heavy coupling and dependency with regard to feasibility considerations: compositionality and composability. These two aspects are outlined in more detail in Table 11.2.

Cyber-Physical Systems are unique in composition as they are inherently heterogeneous in nature throughout, including with regard to essential design requirements [19, 20]. During design considerations, Software Engineers must consider physical requirements in addition to the traditional functional requirements present in common day applications. All aspects of the Cyber-Physical System are codependent, therefore any changes in a design in one will directly impact the other aspects contained within the system. This structure yields most traditional design methodologies useless, since they are normally focusing on a separation of concerns during design implementation [4]. Numerous overlapping among the different system layers is prevalent within the Cyber-Physical system, thereby requiring a unique design approach.

Clearly a more adaptive means to design is necessary to accurately construct an applicable design to support effective CPS design for a number of domains of impact. CPS design can benefit from an automated approach that takes into account complex dependencies, but if this is to be effective it must incorporate an overarching architecture as a guide to develop the KBE system for a particular type of CPS system in question. Such design automation must incorporate all aspects contained within the Cyber-Physical System: computation, physical, and network.

When we use CPS components, it is entirely possible that some components could adapt themselves automatically to the other components in their assembly, which inevitably changes the way in which these CPS-enabled components are designed and manufactured [4]. Real-time adaptability requires a unique system structuring when considering design options. Effective tight integration of these components can be accomplished through an overarching architectural framework. The AIDF was designed with a means for incorporating the union of the computation, networking, and physical sections respectively. Knowledge can be collected from sensors that detect changes within the physical environment to populate the knowledge base. More specifically, the sensors can detect aspects

Table 11.2 Feasibility dependencies of component based systems [4]

Two key CPS topics	
Compositionality	System-level properties can be computed from local properties of components
Composability	Component properties are not changing as a result of interactions with other components

from the environment's temperature, air quality of the current environment, and/or current energy usage among many other possibilities to recommend improvements in the design process for future consideration. Figure 11.3 illustrates the communication among the sensors over the wireless sensor network.

Data obtained by the sensors placed in the physical environment is communicated into the system via a wireless sensor network [19, 21]. While the wireless sensor network is in place at a specific facility, other facilities may also be incorporated within the network. All sensors contained in multiple facilities belong to the same wireless sensor network, thereby increasing the data source obtained from the sensors yielding more accurate proposed adjustments by the logic side of the system.

The given AIDF can be modified to incorporate the needs present within the Cyber-Physical System. We propose a new framework, coined Adaptive Cyber-Physical System Framework (ACPSF). This ACPSF can be adapted for CPS design incorporating a specialized wireless sensor network system.

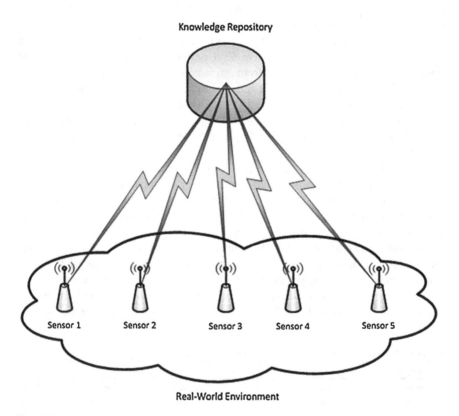

Fig. 11.3 Interactions among sensors among the wireless sensor network present in ACPSF

Adaptive Cyber-Physical System Framework

The central concept to designing the Adaptive Cyber-Physical System Framework is the requirement to handle the design automation concerns for tight integration and emergent properties of a CPS artifact. Special focus must be placed on the need for a design to handle the codependence of the computation, physical, and network system layers. This has been addressed in the Adaptive Cyber-Physical System Framework (ACPSF) based on the design guidelines for KBE engines found in the AIDF. Three similar system engines are also considered by the ACPSF: Knowledge Assimilation Engine, Knowledge Correlation Engine, and Knowledge Justification Engine.

Initially the data is input from two interfaces within the system, one from the physical (or real world) layer and one from knowledge experts (experts in the field currently being studied). These two data sources are utilized conjointly by the Knowledge Assimilation Engine to supply the core dataset to be utilized by the ACPSF. This data is fed directly into the Knowledge Assimilation Engine from the Physical side via the wireless sensor network. The purpose of the Knowledge Assimilation Engine is simply to gather both sets of data and then assimilate them together for the knowledge correlation engine. This data is stored within knowledge repositories, maintained by the Knowledge Assimilation Engine, thereby creating a Knowledge-based System.

After traversing through the Knowledge Assimilation, the data enters the Knowledge Correlation Engine. The Knowledge Correlation Engine accesses the knowledge repositories created and updated by the Knowledge Assimilation Engine through a direct interface to the Knowledge-base portion of the system. Once this data has been accessed by the Knowledge Correlation Engine, proper design analysis and synthesis procedures using the Common KADs approach can be initiated [22]. Several design analysis techniques are utilized in the Knowledge Correlation Engine to achieve a set of potential design recommendations to be made to the system. The design methodologies incorporated into the Knowledge Correlation Engine include modules for FMEA, TRIZ, and FTA among others which are outlined in Table 11.3.

These design analysis techniques can be automated and are valuable for optimizing design considerations for any CPS KBE system. While each design methodology contains benefits for an effective design, we will concentrate on providing some detail on three in particular. By conjoining these three methodologies with several other design analysis techniques we increase the likelihood of delivering an effective and efficient design.

Three modules are described that provide design automation support. Failure mode and effects analysis (FMEA) is an analysis that allows the design to consider potential failures contained in the current design and the impact these failures may have on the system as a whole. Ideally the failures identified here will not contain catastrophic effects. Once this design technique has been applied the design can be adjusted to reduce failures wherever possible. Theory of Inventive Problem

Table 11.3 AIDF design support modules also utilized by ACPSF [2]

Design support modules	Meaning
Axiomatic design theory (ADT)	Provides an automated mechanism for hierarchical decomposition of FR and DP, provides 2 axioms, 11 corollaries, and 23 theorems for the rules base stored in the AI engine block
Theory of inventive problem solving (TRIZ)	Provides an automated mechanism for invention, especially by searching the semantic web for appropriate DPs
Hierarchical multi-layer design (MLH)	Providing an automated mechanism for going from FR to DP to components, calculation of reliability and cost
Quality function deployment (QFD)	Provides an automated mechanism to ensure the customer guidelines are included in the quality of the design
Design structure matrix (DSM)	Provides an automated mechanism to determine component to component interaction
Fault tree analysis (FTA)	Provides an automated mechanism to predict component failures, where the calculations are based heavily on quantitative Boolean operators
Reliability block diagram (RBD)	Provides an automated mechanism to estimate system reliability, where the calculations are based heavily on various engineering equations, such as Mean time to Failure (MTTF)
Failure mode and effects analysis (FMEA)	Provides an automated mechanism to associate weights for each type of failure to assess fault qualitatively and trace root cause
Technology risk factor (TRF)	Provides an automated mechanism to assess individual component risk on a cluster, mainly by associating any given component with a multiplier that affects the DSM
Entropy (ETP)	Provides an automated mechanism to assess level of disorganization in system design during design process
Optical backplane engineering domain (OPT)	Provides an automated mechanism to manipulate domain-specific knowledge for inference, specifically in field of optical backplane engineering
Domain rule support (DRS)	Provides an automated mechanism for domain rule support, in terms of executable rules used by the inference engine
Predicate logic support (PLS)	Provides an automated mechanism for logic support
Algorithmic reasoning support (ARS)	Provides an automated mechanism for miscellaneous algorithmic reasoning support
Fuzzy logic support (FLS)	Provides an automated mechanism for fuzzy logic support
Neural network support (NNS)	Provides an automated mechanism for neural network support
Genetic algorithm support (GAS)	Provides an automated mechanism for genetic algorithm support
Conant transmission support (CTS)	Provides an automated mechanism for component transmission support
Calibrated bayesian support (CBS)	Provides an automated mechanism for calibrated Bayesian support
Data mining support (DMS)	Provides an automated mechanism for data mining support

Solving (TRIZ) is a Russian acronym for an analysis technique that helps to automate innovation and trade-off optimization. Output from the TRIZ process presents the Software Engineer with a contradiction matrix. Contradiction matrices

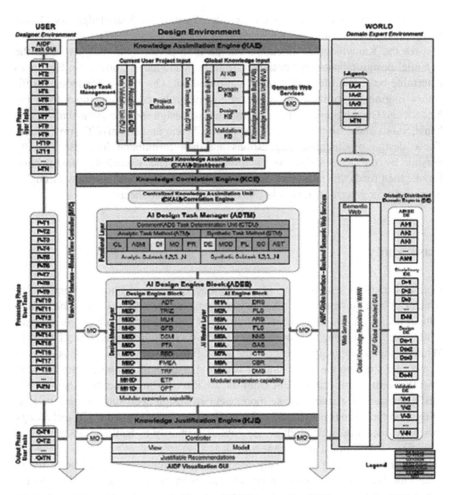

Fig. 11.4 AIDF framework expanded by the ACPSF to develop KBE systems for CPS design automation [2]

simplify the identification of a most appropriate principle to resolve a given contradiction, which helps to resolve current conflicts. Fault tree analysis (FTA) is a technique used in system design that follows a top-down approach to assessing probabilities of failure. There are several benefits to performing this technique. One important benefit is the ability to easily traverse and comprehend the logic and probabilities leading to the top level event. Additional design techniques and modular considerations are applied throughout by the system. These additional techniques and modular considerations are outlined in Table 11.3, which is constructed in relation to the Case Study found in "Case Study". These include the acronyms which are depicted in the Case Study's framework listed as Fig. 11.4. Further elaboration on the applicable design techniques utilized within the ACPSF will be found in a future ACPSF research publication.

All of these design techniques are considered within the Knowledge Correlation Engine and then fed into the Knowledge Justification Engine. The primary function for the Knowledge Justification Engine is to perform an analysis on all the potential design adjustments obtained from the Knowledge Correlation Engine to determine best possible design modification output. Once the Knowledge Justification Engine determines the correct adjustments based on the input from the Knowledge Correlation Engine it forms a list of design modifications that may be beneficial to make for the system to be more efficient and effective for production. These design modification suggestions are then formulated in a user-friendly version by the Knowledge Justification Engine. The Knowledge Justification Engine then feeds this more user-friendly version as output into the view through a graphical user interface, which is utilized by design engineers.

Case Study

Optical backplane engineering is an example of cyber-physical system design which can be automated with a KBE system adapted from the ACPSF. The field of optical backplane engineering is comprised of three distinct technological areas, which are fiber-based, light-wave, and Free-Space Optical Interconnect (FSOI). While it would be possible to apply the ACPSF to any of the three, focus is given in our case study to the FSOI since it is a frontier technology which may stand to gain the most significant benefit from the ACPSF. By implementing the ACPSF, OBIT designers are able to make more accurate design decisions through the risk-mitigation techniques provided by the ACPSF. An implemented FSOI technology can be divided into four modular assemblies primarily to facilitate assembly and servicing of the optical interconnect [23].

Copper replacement has already been the discussion over the course of recent years and has yielded some optical utilization in current practices. Due to the limitations of the current copper infrastructure it is critical that these are replaced with optical solutions to keep up with the increasing demands of data throughput [24, 25]. Given this has a broad impact on the future, it makes it even more ideal to be applied as our case study.

For the modulator-based system shown in Fig. 11.5, the optical interconnect is divided into four functional modules with constituent components:

(1) Beam combination module (BCM) routes the receiving, read-off, and transmitting beams propagating through via interconnect.
(2) Relay module propagates the transmitting beams from one stage to the next.
(3) Chip module performs the electronic-to-optical and optical-to-electronic conversions between the motherboard electronics and the optical interconnect.
(4) Optical power supply (OPS) generates a Two-Dimensional (2D) array of continuous wave (CW) "read-off" beams aligned onto the modulators [26].

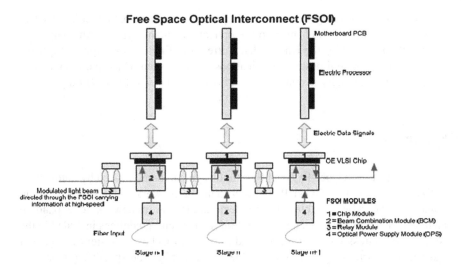

Fig. 11.5 FSOI modular-based system

All the FSOI modular subsystems and components can be gathered from domain experts in the field through the GUI incorporated for data gathering. This data obtained from the domain experts is then stored into the knowledge repositories within the Knowledge Assimilation Engine. The data is stored after being translated into a Web Ontology Language (OWL) readily available to interpretation from the intelligence agents contained within the Knowledge Correlation Engine.

The input obtained from the experts can then be supplemented by that of the physical environment through the use of sensors. We can place sensors within the environment that is utilizing the FSOI in order to monitor the environmental impact, e.g. temperature. Computing equipment must be kept at a stabilized low temperature in order to keep efficient operations of the system. Therefore the temperature sensor readings can provide relevant data feedback into the Knowledge Assimilation Engine via the wireless sensor network. The wireless sensor network communicates all the sensor readings directly into the Knowledge Assimilation Engine based on data throughput on the network from all the monitored environments currently in operation. This input is useful in order to provide design aspects relating to maintaining the physical environment at ideal conditions for operations.

After the Knowledge Assimilation Engine has processed the newly entered data, the data enters into the Knowledge Correlation Engine. Design analysis techniques are performed to determine the best design recommendations based on the feedback obtained from the experts regarding the FSOI modular subsystems and components coupled with the sensor (physical) readings with respect to temperature in the environment. Thus, both environmental data collected from sensors and accumulated domain knowledge acquired from experts are taken into consideration by the ACPSF.

Finally, the Knowledge Justification Engine processes the data in order to select the best potential design changes to the optical backplane system. An analysis of the results provided by the Knowledge Correlation Engine is computed to determine the best potential design modifications to the system. Once these have been selected, a user-friendly display reads and displays them to the GUI output.

Summary

The complexity of Cyber-Physical System design can be addressed by automation utilizing the Adaptive Cyber-Physical System Framework (ACPSF). The broad impact of the AIDF enables its architectural design framework to expand to include Cyber-Physical Design Automation with special guidelines included in the ACPSF to address CPS concerns. Traditional approaches to system design fall short in CPS artifact design, mostly due to the application-specific, robust, and adaptive design methodologies required to obtain feasible and effective designs for the Cyber-Physical System. The ACPSF is able to adequately meet the requirements to improve the CPS design process by introducing intelligent automation methodologies utilizing Web Services and networked knowledge bases. The ACPSF accomplishes this through the three core engines: Knowledge Assimilation Engine, Knowledge Correlation Engine, and Knowledge Justification Engine on the computation side working in close conjunction with the physical and network side. Both human experts populating networked knowledge bases and manufacturing facilities accumulating feedback data with sensors located globally provide valuable design information to the KBE configured by the ACPSF. This knowledge fusion combined with automated analysis and synthesis provides recommendations in order to adequately address the tight integration present in the CPS computation, network, and physical layers respectively.

Future Work

The ACPSF has broad impact across many industries and can be scaled by adding as many modules as needed to configure the best KBE system for a particular application. The complexities of design automation can be systematically addressed with more intellectual control, including but not limited to CPS artifacts needed for energy, medical, transportation, space, and military applications.

References

1. Tanik, U., J. *Artificial Intelligence Design Framework for Optical Backplane Engineering.* Dissertation (2006).
2. Tanik, Urcun, UAB Dissertation, Chair: Gary Grimes.
3. Tanik, Urcun. *Architecting Automated Design Systems.* VDM Verlag, 2008.
4. CPS Steering Group. *Cyber-Physical Systems Executive Summary.* Mar. 2008. Web. 13 Sept. 2012. <http://precise.seas.upenn.edu/events/iccps11/_doc/CPS-Executive-Summary.pdf>.
5. UC Regents. *Cyber-Physical Systems.* <http://cyberphysicalsystems.org/>.
6. Rajkumar, Ragunathan (Raj), et al. "Cyber-Physical Systems: The Next Computing Revolution." *Proceedings of the 47th Design Automation Conference* (2010): ACM Digital Library. Web. 15. Sept.2012
7. Fallah, Yaser, P., et al. "Design of cooperative vehicle safety systems based on tight coupling of communication, computing and physical vehicle dynamics." *Proceedings of the 1st ACM/ IEEE International Conference on Cyber-Physical Systems* (2010): ACM
8. Lee, Edward, A., Seshia, Sanjit, A. "An introductory textbook on cyber-physical systems." *WESE '10: Proceedings of the 2010 Workshop on Embedded Systems Education* (2010): ACM digital Library. Web. 26.Sept.2012
9. Trevino, L., D. E. Paris, and M. Watson, 2005, *A Framework for Integration of IVHM Technologies for Intelligent Integration for Vehicle Management*, NASA Marshall Space Flight Center, Technical Report, Document ID: 20050162249, NTRS: 2005-05-17.
10. Schreiber, Guus, et al. "Knowledge Engineering and Management."Massachusetts Institute of Technology (2000): MIT.
11. Mylopoulos, John, et al. "Building knowledge base management systems." *The VLDB Journal 5: 238* (1996): The VLDB Journal
12. Chung, Lawerence, Cooper, Kendra. "A knowledge-based COTS-aware requirements engineering approach." *Proceedings of the 14th international conference on Software engineering and knowledge engineering* (2002): ACM.
13. Shin, Eun-Hwan. *Estimation Techniques for Low-Cost Inertial Navigation.*2005. University of Calgary. <http://www.ucalgary.ca/engo_webdocs/NES/05.20219.EHShin.pdf>
14. Gevarter, William, B. *An Overview of Artificial Intelligence and Robotics.* 1983. <http:// www.dtic.mil/cgi-bin/GetTRDoc?Location=U2&doc=GetTRDoc.pdf&AD=ADA364025>.
15. Stoilos, G., Stamou, G., Pan, J. Z. 2010. "Fuzzy extensions of OWL: Logical properties and reduction to fuzzy description logics". *International Journal of Approximate Reasoning*, Vol. 51 Issue 6, pp. 656–679.
16. Djelloul, Khalil. "From exponential to almost linear decomposability of finite or infinite trees." *Proceedings of the 2009 ACM symposium on Applied Computing* (2009): ACM.
17. Hazewinkel, Michiel, et al. "Axiomatic method."*Encyclopedia of Mathematics* (2001): Springer
18. Axiomatic Design. *Acclaro DFSS Summary.* <http://www.dfss-software.com/ dfss_summary.asp>
19. Kim, Ji, Eun, Mosse, Daniel. "Generic framework for design, modeling and simulation of cyber-physical systems." SIGBED Review, Vol. 5 Issue 1 (2008): ACM.
20. Kim, Ji-Yeon, et al. "Abstracted CPS model: a model for interworking between physical system and simulator for CPS simulation (WIP)." *Proceedings of the 2012 Symposium on Theory of Modeling and Simulation - DEVS Integrative M&S Symposium* (2012): Society for Computer Simulation International.
21. Ma, Xuan, et al. "A cyber-physical system based framework for motor rehabilitation after stroke." *Proceedings of the 1st International Conference on Wireless Technologies for Humanitarian Relief* (2011): ACM Digital Library. Web. 15.Sept.2012
22. Schreiber, G., H. Akkermans, A. Anjewierden, R. de Hoog, N. Shadbolt, W. Van de Velde, and B. Wielinga. *Knowledge Engineering and Management, The Common- KADS Methodology.* Cambridge, MA: MIT Press, 2000.

23. Kirk, A. G., D. V. Plant, M. H. Ayliffe, M. Châteauneuf, and F. Lacroix, 2003, "Design Rules for Highly Parallel Free-Space Optical Interconnects" *IEEE journal of selected topics in quantum electronics*, Vol. 9, No. 2, pp. 231–245.
24. John H. Glenn Research Center. *Free-Space Optical Interconnect Employing VCSEL Diodes.* Nov. 2009. <http://www.techbriefs.com/component/content/article/5920>.
25. Berger, Chrisoph, et al. *High-density optical interconnects within large-scale systems* 2003. <http://www.zurich.ibm.com/pdf/sys/berger_proc_spie_4942_pp222-235_2003.pdf>.
26. Tanik, U.J., G. J. Grimes, C. J. Sherman, V. P. Gurupur, 2005, "An Intelligent Design Framework for Optical Backplane Engineering," *Journal of Integrated Design and Process Science,* Vol. 9, No. 1, pp.41–53.
27. Zhu, Ting. *Research Interests.* <http://www.cs.binghamton.edu/~tzhu/>.

Chapter 12
Cyber-Physical Ecosystems: App-Centric Software Ecosystems in Cyber-Physical Environments

David E. Robbins and Murat M. Tanik

Introduction

The advent of app-centric software ecosystems is widely known; its most popular manifestations have come in the form of software platforms for mobile devices, such as Apple's iOS [1] or Google's Android [2]. App-centric software ecosystems are also beginning to emerge in Web 3.0 [3] applications, such as pathology image analysis [4] and clinical data management [5]. In general, app-centric software ecosystems are comprised of two elements: a collection of apps, and an environment for the management and execution of those apps. One key element of this environment is a shared data bus between apps.

Simultaneous to the growth in popularity of app-centric software ecosystems, the increasing synergistic effects of tightly coupled physical and computational elements in embedded systems led to the recent introduction of the concept of Cyber-Physical Systems (CPS) [6]. Abstractions previously familiar in software systems fall short when applied to integration of CPS into engineered solutions [7].

To address the need for a new integration framework for CPS, this paper proposes the application of concepts and abstractions from app-centric software ecosystems. This integration of two conceptual frameworks leads to the creation of Cyber-Physical Ecosystems (CPE): app-centric software ecosystems whose apps are CPS, and in which the execution environment is the physical environment of these constituent CPS.

D. E. Robbins (✉)
Department of Electrical and Computer Engineering, University of Alabama at Birmingham, 1150 10th Avenue South, Birmingham, AL 35294, USA
e-mail: robbinsd@uab.edu

M. M. Tanik
Department of Electrical and Computer Engineering, University of Alabama at Birmingham, 1150 10th Avenue South, Birmingham, AL 35249, USA
e-mail: mtanik@uab.edu

S. C. Suh et al. (eds.), *Applied Cyber-Physical Systems*,
DOI: 10.1007/978-1-4614-7336-7_12,
© Springer Science+Business Media New York 2014

Introduction of the concept of CPE begins with a brief background in both CPS and app-centric software ecosystems. Following this, an in depth discussion of integration strategies for the conceptual frameworks provided by app-centric software ecosystems and CPS is given, leading to the description of CPE as an integrated abstraction.

Cyber-Physical Systems

Cyber-physical systems (CPS) are engineered systems designed with an emphasis on extremely tight integration and synergistic effects between the physical and computational elements of the system [6]. The physical elements of a CPS may include sensors and actuators in the traditional sense, such as photocells, physical buttons, motors, and displays or LED's. However, the increasing number of material properties and physical characteristics subject to electrical control allow CPS to modify these properties and characteristics as they pertain to its housing, forming a programmable matter [8].

From a computationally oriented approach, CPS can also be viewed as computational systems with input and output in the physical world. Viewing the elements of a CPS as computational elements allows the potential application of software engineering concepts, such as object-oriented development and high-level process architectures to physical machines, in addition to the logical machines of software.

CPS differentiate themselves from traditional embedded systems by emphasizing a network of computational systems with physical input and output, rather than emphasizing the use of a computational system to drive a single device (e.g. a car or microwave) [7]. That is, a CPS represents not only a single device, but also a collection of devices working in concert. Thus, CPS further extend beyond embedded systems by offering a concept of programmable matter [8], in which individual devices represent atoms, and their interactions with each other and with their environment allow them to assemble into arbitrary physical artifacts.

Within the context of programmable matter, CPS leverage complexity to allow novel applications of computation to the physical world. Complex systems are those that exhibit non-trivial emergent collective behavior and adaptation by processing signals and information [9]. In programmable matter CPS, which represent CPS comprising massive number of CPS atoms, this collective behavior is manifest in the interaction between the individual CPS atoms, while signal and information processing is performed within computational elements of the atoms themselves. Since the atoms are likely to individually have simple behavior, given their necessary small size for most applications [8], their collective interactions represent emergent behavior. Definitions of complexity emphasizing the plasticity of both the roles of individual elements within a system and the network of connections between elements [10] are especially descriptive of programmable matter CPS: the very nature of atomic elements is to be interchangeable.

The effects of increased complexity in both programmable matter and other CPS leads their study to represent a formalization of the general theory of integrating computational and physical systems [11], of which traditional embedded systems represent a subset. Indeed, the underlying theory of controlling physical behaviors using computational elements is surprisingly primitive. Independently considered, we understand well the manipulation of physical properties and behaviors, the execution of computation in electronic computers, and the mathematics of control systems. Despite this understanding, however, the integration of knowledge from these domains into complex systems lacks a theoretical model. Although both the physical and, to a lesser degree, computational elements of a system may be described in detail by theoretical models from their individual domains, an integrated theory deeply elucidating their interaction does not yet exist. Thus, in most applications of embedded systems, a design emphasis is placed either on the physical or computational behaviors of the system, based on the designers modeling background. CPS will transcend these disciplinary shortcomings by providing an integrated view of physical computational systems.

App-Centric Software Ecosystems

App-centric software ecosystems represent a use of software ecosystems as software architecture [12]. In this architecture, the fundamental software module is the app, itself a discrete software system with its own input, output, and internal logic. Apps also represent the primary entity type in the software ecosystem view of the architecture.

Current example's of app-centric software ecosystems are best known in the mobile sector, with the success of Apple's iOS platform [1] and Google's competing entry, Android [2]. However, the success of these platforms has already led to the creation of app stores in a broad set of fields and across a range of products, extending even into the electronic medical records (EMR) domain with the recent SMART Platforms [13]. Web 3.0 [3] based app-centric software ecosystems are also emerging, for pathology image analysis [4] and clinical data management [5].

Architecture of App-Centric Software Ecosystems

Similar to embedded systems and CPS, the theory underlying app-centric software systems lags behind the practice of building them. Despite their widespread use, no formal description of the architecture for app-centric software ecosystems exists.

Fundamentally, an app-centric software architecture represents the composition of a software system and workflow from a number of apps, where each app is a relatively small, single purpose software tool. Across the many implementations of app-centric software ecosystems, two common elements exist: a set of apps, and the environment in which these apps are managed and executed. These two subsystems constitute the architecture of an app-centric software ecosystem.

Apps

As previously mentioned, an app is a small, single purpose software tool. Apps are similar in concept to *modules* or *plugins* in large, single vendor Software-as-a-Service (SaaS) offerings. An app, however, is more independent from its environment than a plugin or module: apps generally have no dependencies other than the basic, general-purpose functionality of the app environment. Nevertheless, apps are sometimes called modules [4], depending on the terminology selected by the ecosystems creator. In general, a module or plugin may be considered an app if it meets these requirements:

(1) Apps have *user* interfaces. That is, apps are designed for interaction with the human users of the system. They may also interact with each other through the app environment.
(2) Apps serve a single or few purposes within their ecosystem (e.g. Email, Calendars, To Do lists).
(3) Apps are discoverable and loadable by the app environment.

These factors also serve to distinguish apps from components and services. In a software module abstraction hierarchy, apps may be considered to exist at a similar level of abstraction to services, while components exist at a lower level.

Comparison Between Apps and Components

Apps are to components as components are to objects: a level of abstraction above. Apps may be composed of components, along with objects, functions and raw procedural code.

While components store the majority of their data, apps store very little of their own data. Rather than storing data internally, apps operate on data that is stored in a shared data bus, provided by the app environment, to facilitate sharing among apps with similar purposes. See the discussion of the app environment for more information.

Unlike components, apps are composed into larger software systems by co-location within an app-environment. Composition of components generally requires configuring a series of interface calls between individual components.

Comparison Between Apps and Services

Apps and services exist at the same level of a software module abstraction hierarchy. However, apps and services take opposite approaches for linking data with executable code. Services generally require the transmission of data to and from the service; in the case of web services this transmission takes place over a data network to remote service hosts. The use of Apps, on the other hand, requires the transmission of the executable code, namely the app, to the execution environment that contains the data.

Both apps and services may be composed of objects, components, functions, and raw procedural code. Unlike apps, services are integrated and composed into larger solutions by chained transformations of a set of data.

App Environment

The app environment of an app-centric software ecosystem provides the resources necessary for discovery, management, and execution of apps. Discovery of apps is often managed through a centralized web service, an app store. The environment's ability to install, configure, and remove apps discovered in the app store constitute its ability to manage the apps it contains. While apps are executed, the environment provides interfaces to shared data storage, external communication (such as network services) and access to the physical devices connected to the app environment, such as cameras, GPS receivers and accelerometers.

In app-centric software architectures, data is shared between apps through a shared data bus, provided by the environment. The use of a shared data bus allows multiple instances of a given type of app to exist within the ecosystem, allowing users to select the instance they prefer to use. For example, Apple's iOS app environment provides access to a calendar database, which apps may use to manage calendar data (e.g. appointments). Users of iOS devices may select from a variety of calendar management apps, based on their preferences, such as a preference for a given user interface.

The shared data bus provided by the app environment also enables apps to react to the activities of other apps in the absence of direct calls from one to another. Returning to the calendar data example in iOS, an app could watch for appointments in the near future, interact with a navigation and GPS apps to determine the time required to travel to the appointment from the devices current location, and trigger an alert or alarm when the user of the device needs to begin traveling to their next appointment. In this example, the user need not enter their appointments directly into the alerting app. Instead, the alerting app reacts to calendar data on the shared data bus.

Notably, nothing in the description of an app-centric software ecosystem's environment constrains the ecosystem's environment to a single device. If the shared data bus is capable of spanning multiple devices, as in a Web 3.0 Resource Description Framework bus [3] or a wireless sensor network, apps on multiple physical devices could still be considered to be part of the same app environment.

Cyber-Physical Ecosystems

Integration of the concepts for app-centric software ecosystems and Cyber-Physical Systems (CPS) leads to the concept of Cyber-Physical Ecosystems (CPE). A Cyber-Physical Ecosystem is an app-centric software ecosystem in which portions of the app environment are physical, rather than logical, entities. In this ecosystem, the primary input and output of each app is cyber-physical in nature: its primary interaction with its environment takes place through sensors and actuators. Conversely, any CPS that meets the criteria necessary for classification as an app may be said to operate in a CPE.

Since CPS may be nested recursively, as in the case of programmable matter, an individual CPS may itself constitute a CPE. This leads to a definition of a CPE as a CPS whose component CPS's meet the criteria for consideration as apps.

Cyber-Physical Ecosystem Integration

The CPE concept provides an integration strategy for CPS, when those CPS meet the criteria for apps. Integration of such CPS into higher-level systems, in a CPE, takes place simply by placing those CPS in the same environment. Notably, this integration strategy emphasizes the functional characteristics of each CPS. Like apps, CPS are integrated into a CPE based on the users needs. In this way, CPE are amenable to semantically driven, information architecture based integration. To discover the particular CPS necessary to achieve a certain functionality in a CPE, a process of decomposition by concept map may be used [14].

Apps and CPS function together in similar fashions: a number of self-similar elements, each with potentially simple behavior, work together to exhibit emergent collective behavior. With this emergent behavior, collections of apps and individual CPS elements become capable of accomplishing complex tasks.

Like most embedded systems, CPS function in response to their sensed environment. Apps represent a software-based equivalent to this behavior of embedded systems: one of the primary characteristics of apps is their ability to react to data on the shared bus in their environment. In CPE, this shared data environment extends to include the physical environment in which the CPS's that make up the CPE operate.

Conclusion

The conceptual frameworks and information architectures of App-centric software ecosystems and cyber-physical systems (CPS) provide complimentary approaches to modeling systems. When integrated, these frameworks combine to become cyber-physical ecosystems (CPE). CPE are CPS whose constituent members, themselves CPS, meet the criteria for consideration as apps. Alternatively, CPE may be viewed as app-centric software ecosystems whose shared data environment includes the physical operating environment of the apps. By taking an app-centric approach to CPS integration, a semantically driven, information architecture based decomposition of a larger CPS into its component CPS becomes possible.

References

1. Apple Inc., "iOS 5," 2012. [Online]. Available: http://www.apple.com/ios/. [Accessed: 02-May-2012].
2. Google, "Android," 2012. [Online]. Available: http://www.android.com/. [Accessed: 02-May-2012].
3. J. Hendler, "Web 3.0 Emerging," *Computer*, vol. 42, no. 1, pp. 111–113, Jan. 2009.
4. J. Almeida et al., "ImageJS: Personalized, participated, pervasive, and reproducible image bioinformatics in the web browser," *Journal of Pathology Informatics*, vol. 3, no. 1, p. 25, 2012.
5. D. E. Robbins, M. M. Tanik, and J. S. Almeida, "Architecture of App-Centric Software Ecosystems for Semi-Structured Clinical Data," in *SDPS 2012*, 2012.
6. National Science Foundation, "Cyber-Physical Systems (CPS)." Washington D.C., pp. 1–12, 2012.
7. E. a. Lee, "Cyber-Physical Systems: Design Challenges," *2008 11th IEEE International Symposium on Object and Component-Oriented Real-Time Distributed Computing (ISORC)*, pp. 363–369, May 2008.
8. J. Campbell, S. Goldstein, and T. Mowry, "Cyber-Physical Systems," Pittsburgh, 2006.
9. M. Mitchell, *Complexity*. Oxford: Oxford University Press, 2009, pp. 12–13.
10. L. Amaral, "Complex systems and networks: challenges and opportunities for chemical and biological engineers," *Chemical Engineering Science*, vol. 59, no. 8–9, pp. 1653–1666, May 2004.
11. W. Wolf, "Cyber-physical Systems," *Computer*, vol. 42, no. 3, pp. 88–89, Mar. 2009.
12. M. Cataldo and J. D. Herbsleb, "Architecting in software ecosystems," in *Proceedings of the Fourth European Conference on Software Architecture Companion Volume - ECSA '10*, 2010, p. 65.
13. K. D. Mandl et al., "The SMART Platform: early experience enabling substitutable applications for electronic health records.," *Journal of the American Medical Informatics Association: JAMIA*, pp. 597–604, Mar. 2012.
14. D. E. Robbins, "Design Space Decomposition Using Concept Maps," University of Alabama at Birmingham, 2011.

Chapter 13
Risk Assessment and Management to Estimate and Improve Hospital Credibility Score of Patient Health Care Quality

Mehmet Sahinoglu and Kenneth Wool

Introduction

The purpose of this chapter is to study how to assess patient-centered health-care quality and as a follow-up, how to mitigate the unwanted risk to a tolerable level, through automated software utilizing game-theoretic risk computing. This chapter overall seeks methods about how to improve patient-centered quality of care in the light of uncertain nationwide health care quality mandate to disseminate and utilize results for the "most bang for the buck". A patient-centered composite 'credibility' or 'satisfaction' score is proposed for the mutual benefit of patients seeking quality care, and hospitals delivering the promised healthcare, and insurance companies facilitating a financially accountable healthcare. Patient-centered quality of care risk assessment and management are inseparable aspects of health care in a hospital, yet both are frequently overlooked. In Alabama State, a 2004 study by the Kaiser Family Foundation found substantial dissatisfaction with the quality of health care. In response to whether they were dissatisfied with the quality of healthcare, 44 % of Latinos, 73 % of Blacks, and 56 % of Whites said "Yes". When asked whether health care has gotten worse in the prior five years prior, 39 % of Latinos, 56 % of Blacks, and 38 % of Whites reported dissatisfaction [1].

Being overly optimistic, and not considering or preparing for possible detrimental events could be severely damaging to both the patient and hospital management. Characterizing and assessing the patient-centered quality of care

M. Sahinoglu (✉)
Informatics Institute, Cybersystems and Information Security, Auburn University at Montgomery(AUM), Montgomery, AL 36124, USA
e-mail: msahinog@aum.edu

K. Wool
Cardiology, Central Alabama Veterans Health Care System, 215 Perry Hill Road, Montgomery, AL, USA
e-mail: drwooke@yahoo.com

S. C. Suh et al. (eds.), *Applied Cyber-Physical Systems*,
DOI: 10.1007/978-1-4614-7336-7_13,
© Springer Science+Business Media New York 2014

(service) risk situation or how to cost-optimize the undesirable risk to a tolerable level within the available budgetary and personnel resources, is not a task one can simply over- or underestimate using a hand calculator. To address this need, the authors will investigate the foundational aspects within an associated automated software tool for cost efficient quantitative risk management. The primary author's innovation, i.e. Risk-O-Meter (RoM), will provide a measurable patient-centered quality of care risk, associated cost, and risk mitigation advice for vulnerabilities and threats associated with automated management of health care quality in a hospital or clinic. The RoM will be demonstrated to assess and enhance quality in the case of an ambulatory or non-ambulatory patient seeking health care at a local hospital. The Quality of Service (QoS) or conversely Risk of Service (RoS) out of a scale of 100 will be estimated [2]. The RoS metric will be followed up by a cost-optimized game-theoretic analysis of how to bring an undesirable risk to a tolerable level by determining what priorities to be taken for which cautionary actions prioritized [3].

The purpose of chapter to study how to assess the quality (of care) which is defined as a measure of the ability of a doctor, hospital or health plan to provide services for individuals and populations that will increase the likelihood of desired health outcomes. These said outcomes are to be consistent with current professional medical knowledge. Good-quality healthcare means doing the right thing at the right time, in the right way, for the right person and getting the best possible results. According to the mantra for the quality improvement movement [4], care should be "safe, effective, patient-centered, timely, efficient and equitable".

To achieve quality improvement, methods should be available to determine "Quality Measures" as the mechanisms used to assign a "quantity" to wellness of care by comparison to a criterion, which in our case constitutes "patient-centered healthcare quality satisfaction" [5]. This chapter aims for these mechanisms through automated software. The chapter content aims to mitigate risk and minimize the risk-mitigating investment costs to achieve goals.

In the healthcare context, the goal of quality improvement strategies is for patients to receive the appropriate care at the appropriate time and place with the appropriate mix of information and available supporting resources. In many cases, healthcare systems are designed in such a way as to be overly cumbersome, fragmented, and indifferent to patients' needs. The patient centered approach is the newest model of many to come down the halls of medical care. The new approach involves a care team, rather than being physician centric, i.e. the pharmacist, primary care doctor, psychologist, pharmacist, dietician, and nurse, seeing multiple patients in a group setting. The co-author's prediction is that this approach may be very useful (that is shared appointments) in certain patients, especially in terms of education efforts.

Quality improvement tools range from those that simply make recommendations but leave decision-making largely in the hands of individual physicians (e.g. practice guidelines) to those that prescribe patterns of care (e.g. critical pathways). Typically, quality improvement efforts are strongly rooted in evidence-based procedures and rely extensively on data collected about processes and outcomes.

This is what the proposed algorithmic software will achieve through an aggregate data collection by running quality risk assessment and risk mitigation using non-subjective risk priority optimization.

Motivation

According to the National Healthcare Quality Report (NHQR) of 2007, the quality of healthcare has only improved at a modest pace; and more importantly the rate of improvement appears to be slowing [5]. Additionally, an important goal of improving healthcare quality is to reduce variations in the quality of care delivery from state to state. Ideally, patients would receive the same high level of quality care regardless of state [6]. The difference between the best and worst performing states, however, can be dramatic as in the NHQR example of diabetes-related hospital admissions (14 times more frequent for worst performing vs. best performing states). Reducing such variation is critical to cost savings. In the case of diabetes-related hospital admissions, had all states been at the level of the best performing states, 39,000 fewer patients would have been admitted with a cost savings of $217 million, according to NHQR. And that is merely for one outcome of just one condition. Another example is the cost attributable to medical errors in lost income, disability, death, and the accrual of additional healthcare costs. That alone is estimated to be $17–$29 billion [7]. Extrapolating from these two examples, potential cost savings of several hundred billion dollars over several years can be envisioned [8]. On the national front, a recent article in the Wall Street Journal highlights the need for hospitals to ensure high quality patient-centered care, particularly in emergency rooms (ER) [9]. The intense and frequently chaotic nature of ER settings, which lack substantial patient data, makes precise patient diagnosis difficult. Anywhere from 37 to 55 % of ER-related malpractice suits stem from these diagnostic errors. It is estimated that such malpractice suits cost over $1 billion in 2009 alone. This proposed algorithmic software, RoM, signifies a critical need for enhancement of patient-centered care quality which will be equally beneficial to hospital quality standings, and nation's rising healthcare costs to avoid misuse, underuse and overuse of equipment and facilities.

So what can be done to improve the delivery of patient-centered quality of care nationally as well as across all states? This is the ultimate goal where automated software is needed and implemented.

One of the primary functionalities of the NHQR is to track improvements in providing safe healthcare. Such tracking is difficult, complex, and must be context-sensitive. The NHQR of 2007 states, "There is still much room for progress in advancing the development of better measurement tools that can help assess whether Americans are obtaining true value in healthcare." What measurement tool is available that can provide this progress?

The answer in part to the above questions is the proposed Risk-o-Meter (RoM). This tool will aid in the improvement of patient-centered quality of care delivery with two critical functionalities:

(1) The RoM will provide an objective, extensible, and adaptable means for tracking the quality of care improvement rate at both the state and national levels.
(2) Further and most importantly, the RoM will identify areas that threaten the delivery of patient-centered quality of care and identify appropriate and cost-effective means to counteract those threats. *No other existing technology provides these unique traits of both cost-savings and improvement of care. The existing ones all fall short of these qualifications* [10].

It is increasingly vital that hospital physicians, other clinicians and auxiliary personnel understand the healthcare system's quality requirements so that they can advocate effectively for their patients and actively assist in health system improvement efforts. The input set of the RoM is an input diagnostic questionnaire designed to measure quantitative attitudes among medical professionals, both clinicians and clinicians-in-training and auxiliary personnel such as nurses, and pharmacists and others, towards aspects of quality of medical practice associated with managed care. This detailed, yet unobtrusive information-gathering quality-control questionnaire includes close-ended opinion statements that could propel changes in the healthcare system, the involvement of alternative health plans, and effective techniques for managing the care of patients and populations. This study also plans to make the algorithmic survey available to University of Alabama at Birmingham (UAB, which controls the Baptist Health System) medical school faculty as a positive instructional tool to educate students, and shape operational attitudes and opinions about the patient-centered healthcare system in which they will eventually practice.

A number of institutions have established their own Health Assessments such as the Mayo Clinic and Tufts Health Care Institute's Online Content Pre/Post Assessment. In contrast to the existing assessments, this paper focus is to provide a generic assessment of a patient's quality of healthcare, once that patient is out of the hospital where he/she was supposed to have been treated to his/her full satisfaction. Interviews with commercial healthcare corporations indicate that there is no such dynamic and interactive tool on the market being used for this purpose [10].

In summary, the study provides healthcare executives and decision makers with an easy, objective, quantitative "patient-centered quality of care risk" assessment and management tool, RoM. In addition to providing an assessment of IT resources vulnerabilities, the RoM offers an objective mitigation advice list in the form of specific recommendations and dollar-based figures about how to enhance quality. Therefore, the RoM is a unique tool that offers healthcare decision makers an innovative alternative in terms of assessing the quality of patient care in a hospital setting. The tool provides specific, practical advice to mitigate the identified vulnerabilities. It also provides a mechanism for the allocation of funds with dollar figures and priority orders to mitigate risk [11]. Working recursively, RoM users can see how much they have lowered their risk so as to take further countermeasures to recursively reduce the risk. This signifies a nonstop 24/7 surveillance and unobtrusive information gathering activity from the actual patients who have entered their hospital portal to take RoM's "satisfaction questionnaire." Therefore, we need a patient-centered quality of care risk assessment device in a non-ambulatory hospital setting, provided the target risk we are after is numerically measurable and improvable in terms of numbers, rather than just qualitative attributes which cannot translate to dollars and cents. Note that *quality measures* are defined as mechanisms used to assign a "quantity" to care, not to append as descriptive adjectives.

Context and Methodology

On top of providing an assessment of IT resource vulnerabilities, the Risk-O-Meter provides an objective mitigation advice list in the form of specific recommendations and dollar figures. The RoM is a unique tool that offers healthcare decision makers an innovative alternative in terms of assessing the degree of Quality of Service (QoS) improvement needed. Based on stakeholder responses, the said RoM as automated software identifies systemic (thorough but specific) vulnerabilities. Maintaining the quality level of patient care at hospitals cannot be accurately accomplished without a risk assessment first and then a risk management of smaller healthcare subsystems, such as smaller pieces of a puzzle, constituting the larger system. The RoM will greatly facilitate conducting an accurate and thorough assessment of the potential risks and vulnerabilities of hospital patient-centered healthcare utilizing the following exhaustive list of vulnerability factors:

(1) Admissions, Billing and Accounting
(2) Hospital Support Services
(3) Outpatients and Daily Visits
(4) Inpatients
(5) Surgery
(6) Emergency Room (Services)
(7) Radiology
(8) Central (all purpose) Labs
(9) IT Resources
(10) Physicians and Interns
(11) Nurses and Auxiliary Personnel
(12) Pharmacy.

Unlike other risk indices that portray risk in terms of a subjective, qualitative high–medium–low scale, the RoM tool offers an objective, quantitative means to identify risks and vulnerabilities. The RoM tool will thus greatly enhance the ability of healthcare executives, decision makers, healthcare insurance providers and IT professionals to maintain patient-centered QoS in a hospital ambience.

What-if questions about how to bring the undesirable Risk-of-Service (RoS) factor as the complement of QoS down to a tolerable percentage will follow. These will be resolved with a roadmap of guidance and a cost effective financial recursive feedback. RoM can also work for the hospital before launching a new enterprise to tailor it (note the Comparative Effectiveness Portfolio) when it is most malleable, so that risks are avoided whenever possible. This entails cost-benefit analyses, risk identification, and assessment with further strategy evaluation through recursive risk management and feedbacks on a continual basis.

The proposed method will also make critical check-listing within hospital healthcare and their follow ups possible and easier than by other non-digitized methods. Dr. Atul Gawande's Checklist Manifesto emphasizes this habit as done in airlines (e.g.: US Air pilot "Sully" Sullenberger used such a procedural checklist in landing on the Hudson River in Jan 2009) and other settings having complex procedures [12]. Dr. Gawande's key message is that the volume and complexity of knowledge today has exceeded any single individual's ability to manage it consistently without error despite material advances in technology, increased training, and super-specialization of functions and responsibilities. Despite demonstrating that checklists produce results, there is widely accepted resistance to their use because our jobs are either too complex to be reduced to a checklist, or because checklists are too rigid and don't force us to look up and think ahead. Yet such a checklist is needed in a complex environment where routine matters that are easily overlooked under the strain of more pressing matters overwhelm people. The RoM software is a scientific methodology and soft technology to get checklists done systematically without having to recall or memorize them one by one, infeasible to do.

By developing and implementing a process checklist for critical processes and decisions regarding a patient's hospital care as ideally described by the books, a disciplined adherence to essential procedures—by checking them off a list—can prevent potentially fatal mistakes and corner cutting. This is what the proposed RoM aims to do by assessing the lack of hospital service quality regarding patient-centered care and by making sure checklists are duly met. Moreover, the proposed study advances planning with a definitive roadmap via a game-theoretical, cost-cutting, and resource-minimization algorithm that is computationally intensive. This process can be performed solely in an automated software engineered environment. Within all these avenues, RoM will guide and help identify the relevant risks relative to each other and work to optimally minimize them to ensure the success of hospital management. This algorithmic software proposed will increase quality of healthcare by proper assessment and mitigation of risk in patient-care using a digital technology through an automated hands-off and objective (not subjective or haphazard or convenient to prove one's opinions) software tool. The advantages are plentiful, but require properly collected authentic and aggregate (composite) patient data analysis. The ultimate goal is to help reach a hospital patient-centered culture of best practices to improve healthcare. Best practices are the most current patient care interventions, which result in the best patient outcomes and minimize patient risk of death or complications to benefit all sides.

Innovation

Quantitative methods are widely employed in healthcare management areas such as forecasting, decision making, scheduling, productivity, resource allocation, supply chain and inventory management, quality control, and project management. Yet when it comes to risk assessment and mitigation of Health IT system risks, qualitative methods currently predominate. The qualitative approaches may be somehow adequate except for periods when budgetary resources are scarce such as during current economic times (2007–present). Consequently, one does not know how to prioritize risks without following an objective computerized plan about how to frugally meet the demands when dealing with only pure adjectives (bad, medium or good). A literature search using the term "quantitative methods for healthcare IT" turned up infrequently. Most of the literature uncovered dealt with qualitative methods, thus showing how little has been done to employ quantitative methods in healthcare IT risk assessment and management [13]. There are very few books and research papers on the topic of quantitative methods in healthcare management [14, 15]. The RoM tool therefore is unique in applying a more rigorous and objective quantitative approach to patient-centered healthcare risk assessment and financial management. There are some new books which address quantitative notions such as "Risk and Exposure Assessment" [Chap. 9, 16] similar to what is proposed here but in a different context. The referenced authors' probability of risk corresponds to a cross-product of vulnerability (=hazard) and threat (=exposure) probabilities. Once treated with a dose of countermeasure, we end up with a residual risk, a concept which in the same book is cited as "precautionary principle". Similarly, "consequence weight" in the cited "Public Health Foundations, Concepts and Practice" by Andresen and Bouldin corresponds to RoM's criticality factor (0.0–1.0) in this proposal where the highest criticality takes on a value of perfect 1.0 such as in a nuclear plant meltdown that happened recently in Japan. A college central computer may have a criticality of 0.4–0.5 [17] whereas a printer may have a weight around 0.2, if not crucial to the business at the specific time.

Risk assessment methods are typically are classified as conventionally qualitative [18–20], newly quantitative [2, 3, 21–23], and also hybrid [2, 3, 24, 25]. The RoM tool uses a quantitative approach for software assurance (the confidence in being free from intentional or accidental vulnerabilities) to determine and manage patient-centered quality of care risk and has the advantage of being objective in terms of dollar figure allocation of mitigation resources. Unfortunately, there is a widespread reluctance to apply quantitative methods [26, 27]. Despite these advantages, decision makers tend to lean toward descriptive risk assessments because they are easy to use and have less rigorous input data requirements. One primary reason is the difficulty in collecting trustworthy data regarding quality breaches elevating risk [28]. A well-known management proverb says that you *can* quantify risk: "What is measured is managed" [29]. The practicality of the proposed method relies crucially on the validity and reliability of the information source for input aggregate data received from the patients.

Approach

Hospital or patient-care centers should be equipped to have "wellness-scores" akin to those of individuals' financial "credit scores" with a list of advisory guidance on which to countermeasure to use improving these risk indicators. In the event of a patient-care center or hospital scoring higher than a standard risk percentage (like a standard or threshold patient-care satisfaction score) after activating and implementing the proposed RoM; the healthcare insurance provider will be authorized to send a warning to the said center to get its act together and remediate or else face the consequences of elevated premiums for their customers (patients). This crucial issue has been recently in the news where WSJ had headlines on its Marketplace section on May 16, 2011, *Wellpoint Shakes Up Hospital Payments* [30]. The article begins with the paragraph, *Wellpoint Inc. is raising the stakes for reimbursing about* 1,500 *hospitals across the country, cutting off annual payment increases if they fail to deliver on the big health insurer's definition of quality patient care.*

To circumvent these universally recognized problems, and hence deliver scientifically objective automated software for risk assessment and risk remediation to serve common purpose, the paper entails the use of RoM. The said software tool will function as a most effective guide to advise the hospital management on how to take countermeasure actions indicated by a cost-optimal game-theoretical algorithm following a risk calculation. The RoM design provides the means in a quantitative manner that is vital in the risk assessment world. Figure 13.1 illustrates the constants in the RoM software as the utility cost (dollar asset) and criticality constant. Figure 13.2 shows the tree diagram where the probabilistic inputs are vulnerability, threat, and lack of countermeasure all valued 0–1.

Risk is generally defined as the likelihood of the occurrence of an event. However, to be able to identify not only the likelihood of the event, but also its impact, we utilize the following definition of risk. Generally speaking, risk is the product of likelihood and impact,

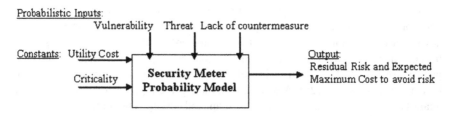

Fig. 13.1 Risk-O-Meter model of probabilistic, deterministic inputs, and calculated outputs

Fig. 13.2 General-purpose
tree diagram (V-branches, T-
twigs, LCM-limbs) for the
RoM software

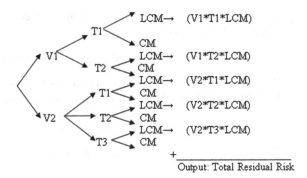

Output: Total Residual Risk

$$Risk = Likelihood * Impact \qquad (13.1)$$

For example, the measure of an Information Technology risk is the product of threat, vulnerability and asset cost:

$$Risk\,(\$) = Threat * Vulnerability * Asset\,(\$) \qquad (13.2)$$

where, vulnerability (equivalent to an ecological component or asset that can become a weakness if exploited and/or misused) refers to the likelihood, and threat (such as an ecological stressor) on the other hand refers to the impact of occurrence, as in Eq. (13.1). The Certified Information Systems Auditor (CISA) Review Manual, 2006, provides the following definition of risk management. Risk management is the process of identifying vulnerabilities and threats to the information resources used by an organization in achieving business objectives, and deciding what countermeasures, if any, to take in reducing risk to an acceptable level, based on the value of the information resource to the organization.

There are two things in this definition that may need some clarification. First, the *process* of risk management is an ongoing iterative process. It must be repeated indefinitely. The business environment is constantly changing and new threats and vulnerabilities emerge every day. Second, the choice of countermeasures (controls) used to manage risks must strike a balance between productivity, cost, effectiveness of the countermeasure, and the value of the informational asset being protected. The *residual risks (RR)*, i.e. the risk remaining after risk treatment decisions have been taken, should be estimated to ensure that sufficient protection is achieved. If the residual risk is unacceptable, the risk treatment process should be re-iterated. Here is where many private entities differentiate between internal costs, costs they must reasonably be expected to pass along to their customers in the pricing of their goods and services, and external costs, those they can pass along to the general public and taxpayers. Introducing the cautionary measures, risk metric is reduced by the probability of countermeasure (CM) action. If for instance, CM probability is perfect (100 %), then the Lack of Countermeasure (LCM) is $1-CM = 0$, reducing the Residual Risk to a merely nonexistent quantity. Residual risk (RR) is a probability between 0 (perfect countermeasure available) and 1.0 (no countermeasure).

$$Residual\,Risk(\$) = Risk\,(\$) * Lack\,of\,Countermeasure \qquad (13.3)$$

The game-theoretic set of equations for risk management follows:
MIN COLLOSS (0<Column loss<1), subject to (see Figs. 13.4, 13.6):

$$1cm11 > 0.765 \qquad (13.4)$$

$$1cm12 > 0.61 \qquad (13.5)$$

$$1cm21 > 0.61 \qquad (13.6)$$

$$1cm22 > 0.385 \qquad (13.7)$$

$$1cm23 > 0.465 \qquad (13.8)$$

$$1cm31 > 0.775 \qquad (13.9)$$

$$1cm32 > 0.725 \qquad (13.10)$$

$$1cm41 > 0.55 \qquad (13.11)$$

$$1cm42 > 0.545 \qquad (13.12)$$

$$1cm43 > 0.525 \qquad (13.13)$$

$$1cm51 > 0.61 \qquad (13.14)$$

$$1cm52 > 0.67 \qquad (13.15)$$

$$1cm61 > 0.33 \qquad (13.16)$$

$$1cm62 > 0.665 \qquad (13.17)$$

$$0.090077cm11 - 1Colloss < 0 \qquad (13.18)$$

$$0.078754cm12 - 1Colloss < 0 \qquad (13.19)$$

$$0.076336cm21 - 1Colloss < 0 \qquad (13.20)$$

$$0.065728cm22 - 1Colloss < 0 \qquad (13.21)$$

$$0.065728cm23 - 1Colloss < 0 \qquad (13.22)$$

$$0.083834cm31 - 1Colloss < 0 \qquad (13.23)$$

$$0.082999cm32 - 1Colloss < 0 \qquad (13.24)$$

$$0.05495cm41 - 1Colloss < 0 \qquad (13.25)$$

$$0.045216cm42 - 1Colloss < 0 \qquad (13.26)$$

$$0.056677cm43 - 1Colloss < 0 \qquad (13.27)$$

$$0.09cm51 - 1Colloss < 0 \qquad (13.28)$$

$$0.09cm52 - 1Colloss < 0 \qquad (13.29)$$

$$0.065192cm61 - 1Colloss < 0 \qquad (13.30)$$

$$0.050692cm62 - 1Colloss < 0 \qquad (13.31)$$

$$0.090077cm11 + 0.078754cm12 + 0.076336cm21 + 0.065728cm22$$
$$+0.065728cm23 + 0.083834cm31 + 0.082999cm32 + 0.05495cm41$$
$$+0.045216cm42 + 0.056677cm43 + 0.09cm51 + 0.09cm52 + 0.065192cm61$$
$$+0.050692cm62 > 0.65$$

$$(13.32)$$

Optimal Solution See columns 3 and 5 in Fig. 13.4 to compare.

$CM_{11} = 0.7795$, $CM_{12} = 0.61$, $CM_{21} = 0.61$, $CM_{22} = 0.385$, $CM_{23} = 0.465$, $CM_{31} = 0.775$, $CM_{32} = 0.725$, $CM_{41} = 0.55$, $CM_{42} = 0.545$, $CM_{43} = 0.525$, $CM_{51} = 0.61$, $CM_{52} = 0.67$, $CM_{61} = 0.33$, $CM_{62} = 0.665$.

Important to note that there may be other alternative solutions to satisfy the constraints. One generated above by the RoM algorithm will be the least costly due to the least amount of percentage sum of changes. See Fig. 13.3 an alternative solution, which amounts to % change of $100[(0.802214 - 0.61) + (0.821968 - 0.775) + (0.830238 - 0.725) + (0.765655 - 0.61) + (0.765655 - 0.67)] = 100[0.59573] = 59.57$ %, which is more than the RoM's least sum change: 53.75 %.

The game-theory application software stabilized this lack of equilibrium with mixed strategy solution. This provides a list of countermeasure probabilities, $CM_{11} = 0.7795,..., CM_{51} = 0.8995,..., CM_{62} = 0.665$. This is the optimal mixed strategy for Defense to minimize its expected loss while Offense maximizes its gain. There is no better game plan at equilibrium by altering CM_{ij}. The author also experimented with Nash equilibrium mixed strategy of probabilities, but the present Neumann approach with mixed strategy generated the scientifically optimal results by Sahinoglu et al. [31].

Next, we plan to apply a scenario to the patient-centered healthcare service at a hospital setting with their set of vulnerabilities, threats and countermeasures as in Fig. 13.6. These classifications of vulnerability-threat-countermeasure are specified by hospital dynamics in relation to their managed patient care implementations. The ultimate purpose is to cost-optimize and prioritize the precautions needed to meet hospital care check-list and quality requirements. This set of actions will improve patients' healthcare by assessing the quantitative risk with a roadmap of what-to-do list at what price and which priority to minimize the risk accrued during the hospital care of the visiting patient. These said goals are

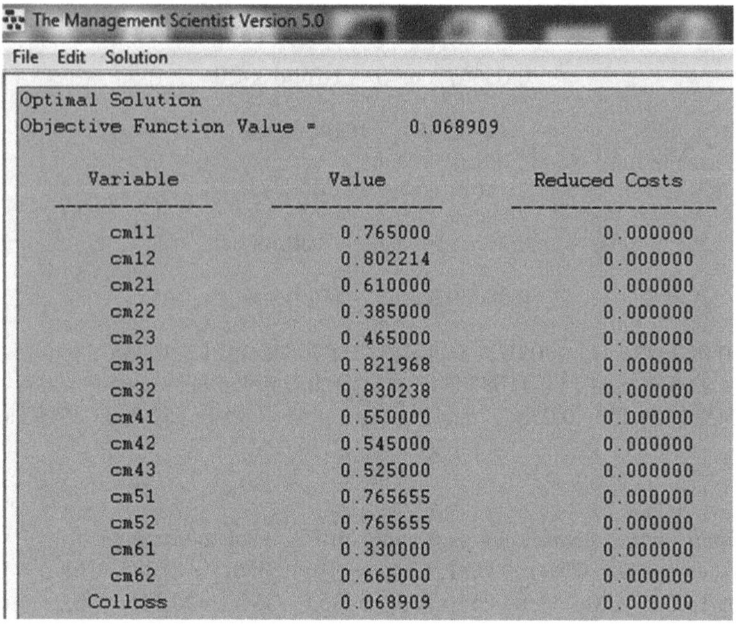

```
🔅 The Management Scientist Version 5.0
File   Edit   Solution

Optimal Solution
Objective Function Value =        0.068909

        Variable              Value              Reduced Costs
     ---------------      ---------------       ---------------
            cm11            0.765000               0.000000
            cm12            0.802214               0.000000
            cm21            0.610000               0.000000
            cm22            0.385000               0.000000
            cm23            0.465000               0.000000
            cm31            0.821968               0.000000
            cm32            0.830238               0.000000
            cm41            0.550000               0.000000
            cm42            0.545000               0.000000
            cm43            0.525000               0.000000
            cm51            0.765655               0.000000
            cm52            0.765655               0.000000
            cm61            0.330000               0.000000
            cm62            0.665000               0.000000
          Colloss           0.068909               0.000000
```

Fig. 13.3 Alternative solution generated by management scientist

actually brainstormed daily, and contemplated nonstop by the hospital administrators who wish to improve conditions, but not readily expressed or delineated to perform in a cohesive manner due to lack of an automated software. RoM realizes the execution of what seems to be intangible goals to a tangible solution so that subjective reasoning is replaced by an objective algorithm for the common good of both patients and administrators, and medical personnel of the hospital. A sample study is drafted starting with Fig. 13.6's tree diagram and a detailed action plan is advised as outlined in Fig. 13.5. The computationally-intensive automated software tool, i.e. Risk-O-Meter will process the diagnostic cognitive (verbal or categorical) and evidential-experiential (numerical) confidence data. Figure 13.4 will verify Fig. 13.5 using 1 M simulations with satisfactory results.

Once the Risk-O-Meter is processed with the input data (the entirety of input test data pending for the main grant if awarded), the assessment output in Fig. 13.3 will need to be interpreted. Using Fig. 13.4 and the detailed Fig. 13.3, to improve the entire operations by mitigating from 39.82 to 35 %, one needs to implement the first-prioritized three counts of recommended 'Countermeasure' actions. (1) Increase the CM capacity for the vulnerability of "Inpatients" and its related threat "Hospital Infections and Insufficient Hygiene and Sanitation" from the current 76.5 to 77.95 % for a performance improvement of 1.45 %, (2) Increase the CM capacity for the vulnerability of "Central Laboratories" and its related threat "Laboratory Personnel Staffing" from the current 61 to 89.96 % for a performance improvement of 28.96 %, (3) Increase the CM capacity for the vulnerability of

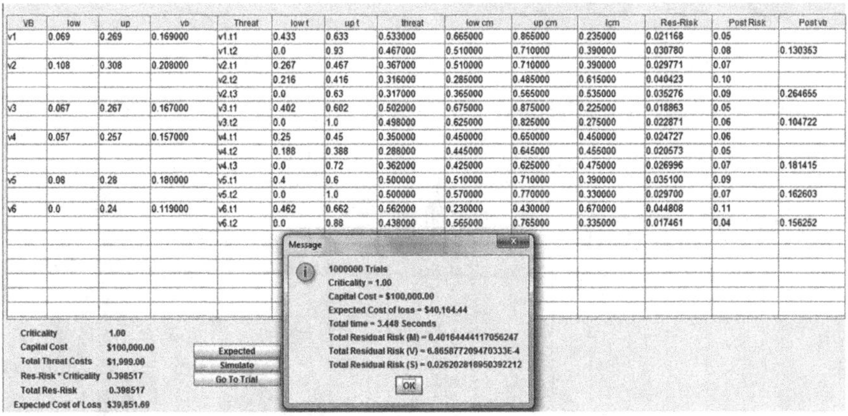

VB	low	up	vb	Threat	low t	up t	threat	low cm	up cm	lcm	Res-Risk	Post Risk	Post vb
v1	0.069	0.269	0.169000	v1.t1	0.433	0.633	0.533000	0.665000	0.865000	0.235000	0.021168	0.05	
				v1.t2	0.0	0.93	0.467000	0.510000	0.710000	0.390000	0.030780	0.08	0.130363
v2	0.108	0.308	0.208000	v2.t1	0.267	0.467	0.367000	0.510000	0.710000	0.390000	0.029771	0.07	
				v2.t2	0.216	0.416	0.316000	0.285000	0.485000	0.615000	0.040423	0.10	
				v2.t3	0.0	0.63	0.317000	0.365000	0.565000	0.535000	0.035276	0.09	0.264655
v3	0.067	0.267	0.167000	v3.t1	0.402	0.602	0.502000	0.675000	0.875000	0.225000	0.018863	0.05	
				v3.t2	0.0	1.0	0.498000	0.625000	0.825000	0.275000	0.022871	0.06	0.104722
v4	0.057	0.257	0.157000	v4.t1	0.25	0.45	0.350000	0.450000	0.650000	0.450000	0.024727	0.06	
				v4.t2	0.188	0.388	0.288000	0.445000	0.645000	0.455000	0.020573	0.05	
				v4.t3	0.0	0.72	0.362000	0.425000	0.625000	0.475000	0.026996	0.07	0.181415
v5	0.08	0.28	0.180000	v5.t1	0.4	0.6	0.500000	0.510000	0.710000	0.390000	0.035100	0.09	
				v5.t2	0.0	1.0	0.500000	0.570000	0.770000	0.330000	0.029700	0.07	0.162603
v6	0.0	0.24	0.119000	v6.t1	0.462	0.662	0.562000	0.230000	0.430000	0.670000	0.044808	0.11	
				v6.t2	0.0	0.88	0.438000	0.565000	0.765000	0.335000	0.017461	0.04	0.156252

Criticality 1.00
Capital Cost $100,000.00
Total Threat Costs $1,999.00
Res-Risk * Criticality 0.398517
Total Res-Risk 0.398517
Expected Cost of Loss $39,851.69

[Expected] [Simulate] [Go To Trial]

Message
1000000 Trials
Criticality = 1.00
Capital Cost = $100,000.00
Expected Cost of loss = $40,164.44
Total time = 3.448 Seconds
Total Residual Risk (M) = 0.40164444117056247
Total Residual Risk (V) = 6.865877209470333E-4
Total Residual Risk (S) = 0.026202818950392212
[OK]

Fig. 13.4 1 M Monte Carlo simulation runs give mean (M) = 0.40 (expected = 0.3985) and standard deviation (S) = 0.026

"Central Laboratories" and its related threat "Patient records" from the current 67 to 89.96 % for a performance improvement of 22.96 %. As indicated in 3, these actions are selected to be the most cost-saving countermeasures, which point out to a total investment of $478.43. This is advised for the breakeven cost of $8.96 per each 1 % improvement. That is, total change of 53.37 % times $8.96 per 1 % = $478.43. The next step by RoM entails carrying on with the optimization to a tolerable percentage once the services are provided. Hospital may lower to 30 % with a 5 % improvement compared to the current 35 % if the budget permits for more services. See a linear system of equations used towards game-theoretic risk computing, and pertinent risk expressions, as shown in "Approach".

The specific objective of this paper is to plan to test and evaluate the RoM, a quantitative risk management tool, in both rural and urban hospital settings and disseminate results and offer feedback once the software is applied for eliciting field data. This process will enable hospital patient-centered quality of healthcare measurement planning, utilizing a definitive roadmap via a game-theoretical cost-cutting and resource-minimization algorithm that is computational intensive. In essence, the RoM software will guide and help identify the relevant risks relative to each other and work to minimize them as optimally as possible to ensure the success of the hospital management. This effort is for both sides of the isle (care seekers and care givers) in trying hard to reach optimal quality. The advantages are plentiful versus the small price of eliciting proper input data (Fig. 13.6).

Validation of this proposal will be accomplished via recursive feedbacks of the RoM algorithm which allows the users in real-time to reconsider the hospitals' varying risks and precautions. The hospital can undertake a review every 6 moths based on the aggregate data by the patients whether in actuality the new improved goal from an earlier undesirable risk level has been met.

The model will effectively develop a monitoring capacity for the quality fulfillment of hospital managed care check-lists prior to fulfilling the patient care

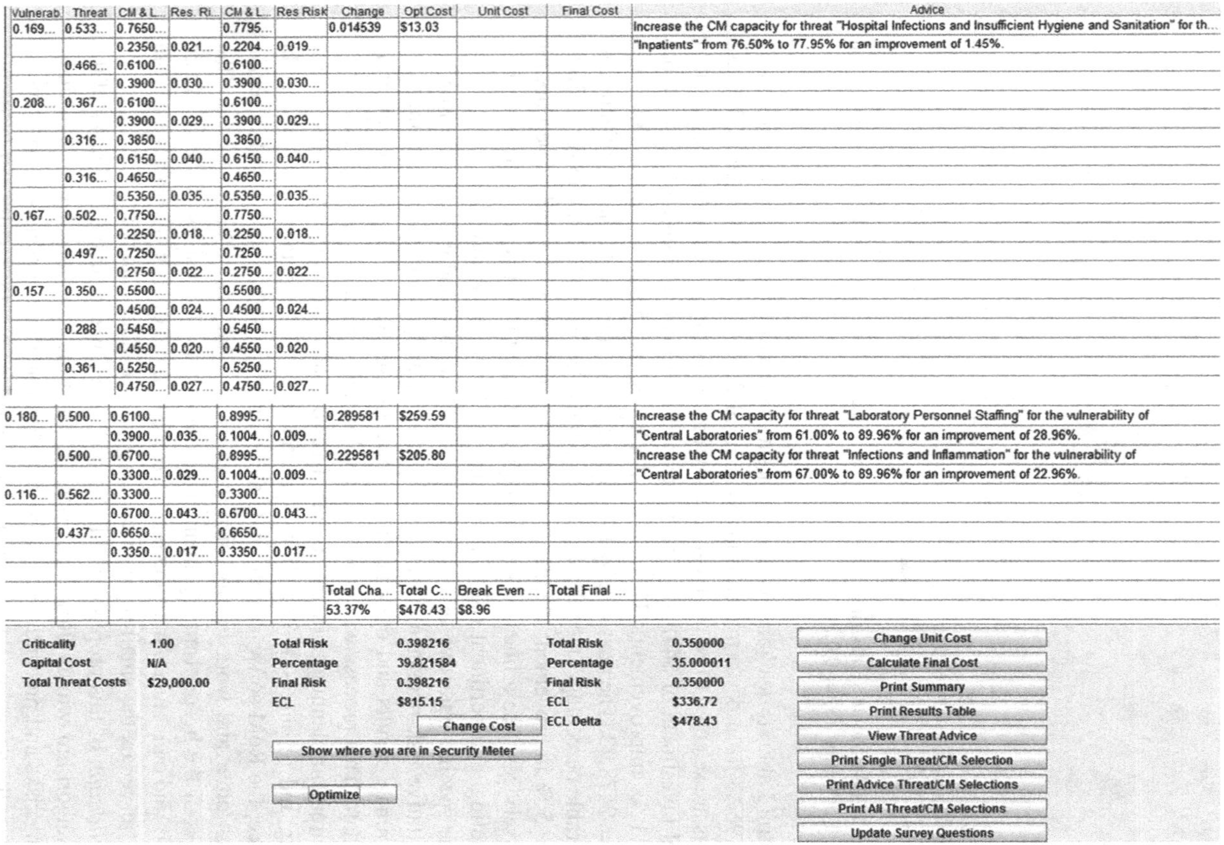

Vulnerab	Threat	CM & L	Res. Ri	CM & L	Res Risk	Change	Opt Cost	Unit Cost	Final Cost	Advice
0.169	0.533	0.7650	0.021	0.7795	0.019	0.014539	$13.03			Increase the CM capacity for threat "Hospital Infections and Insufficient Hygiene and Sanitation" for th…
		0.2350		0.2204	0.019					"Inpatients" from 76.50% to 77.95% for an improvement of 1.45%.
0.466		0.6100	0.030	0.6100	0.030					
		0.3900		0.3900	0.030					
0.367		0.6100	0.029	0.6100	0.029					
		0.3900		0.3900	0.029					
	0.316	0.3850		0.3850						
		0.6150	0.040	0.6150	0.040					
	0.316	0.4650		0.4650						
		0.5350	0.035	0.5350	0.035					
0.167	0.502	0.7750	0.018	0.7750	0.018					
		0.2250		0.2250	0.018					
	0.497	0.7250	0.022	0.7250	0.022					
		0.2750		0.2750	0.022					
0.157	0.350	0.5600	0.024	0.5600	0.024					
		0.4500		0.4500	0.024					
	0.288	0.5450		0.5450						
		0.4550	0.020	0.4550	0.020					
	0.361	0.5250		0.5250						
		0.4750	0.027	0.4750	0.027					
0.180	0.500	0.6100		0.8995		0.289581	$259.59			Increase the CM capacity for threat "Laboratory Personnel Staffing" for the vulnerability of
		0.3900	0.035	0.1004	0.009					"Central Laboratories" from 61.00% to 89.96% for an improvement of 28.96%.
	0.500	0.6700		0.8995		0.229581	$205.80			Increase the CM capacity for threat "Infections and Inflammation" for the vulnerability of
		0.3300	0.029	0.1004	0.009					"Central Laboratories" from 67.00% to 89.96% for an improvement of 22.96%.
0.116	0.562	0.3300		0.3300						
		0.6700	0.043	0.6700	0.043					
	0.437	0.6650		0.6650						
		0.3350	0.017	0.3350	0.017					

	Total Cha…	Total C.	Break Even…	Total Final …
	53.37%	$478.43	$8.96	

Criticality	1.00		Total Risk	0.398216		Total Risk	0.350000
Capital Cost	N/A		Percentage	39.821584		Percentage	35.000011
Total Threat Costs	$29,000.00		Final Risk	0.398216		Final Risk	0.350000
			ECL	$815.15		ECL	$336.72
						ECL Delta	$478.43

Change Cost

Show where you are in Security Meter

Optimize

Change Unit Cost
Calculate Final Cost
Print Summary
Print Results Table
View Threat Advice
Print Single Threat/CM Selection
Print Advice Threat/CM Selections
Print All Threat/CM Selections
Update Survey Questions

Fig. 13.5 Risk-O-Meter assessment (39.82 %) and management plan

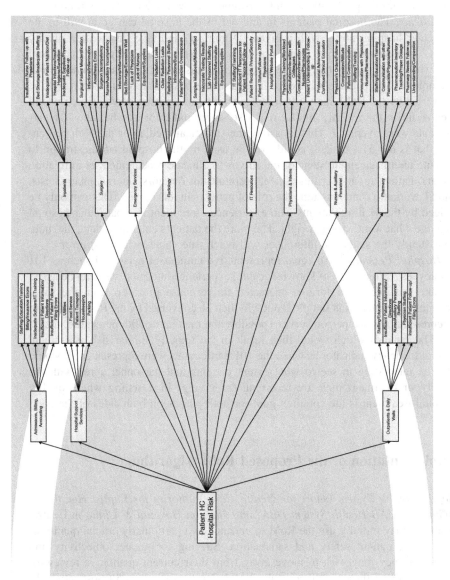

Fig. 13.6 RoM: hospital patient healthcare quality tree diagram

satisfaction. The RoM implementation of patient-centered quality of healthcare improvement through monitoring of the hospital check-list quality mandates as derived from the patient aggregate data will create a model for two local largest Central Alabama hospitals selected for a pilot study. RoM survey results will be obtained from hospital patients and personnel and be implemented as follows.

Patients. The hospital patients in and out of the hospital can go at will to the particular hospital portal with a given ID number, and take this written quality survey questionnaire's first stage (not the second stage cost-optimal management part which is primarily relevant to hospital administrators and stake holders). For example, if 101 patients who were treated at the hospital took it, then it will be converted by the RoM analyst to a representative one-person survey which would then result in the roughly similar risk assessment indicator as when the 101 participants were averaged. This (possibly the statistical median or the 50th percentile, that is the 51st ranked) could then be used by the hospital administration to execute the management stage of the survey to allocate procurement or mitigation dollars. Patient input is critical since, hospital staff (doctors, interns, pharmacists, and IT workers) cannot assume the role of patients in this survey. That can only be judged by the visiting patients who experienced treatment at a particular hospital to judge what went wrong or right. The more the patients enter their data, the more consistently the statistical inferences will reach true values with least error.

Hospital Personnel. An equal representative number of personnel such as 101 employees from a cadre of knowledgeable, experienced hospital staff can be asked to take this survey to form a characteristic response portfolio from the same hospital personnel. This sample input will then be converted by the RoM analyst to a representative one person survey (possibly the median or 50th percentile, that is the 51st ranked) which would then result in the roughly similar risk assessment central tendency indicator as when the 101 participants were represented. It would also be interesting to see to what extent patients and personnel agree with the quality of treatment they are receiving (or giving) in a setting where quality hospital treatment is the common goal for both parties on both sides of the isle.

Implementation of the Proposed RoM Algorithm

Baptist Health System (short for Health Care Authority for Baptist Health- an Affiliate of UAB Health System DBA) and Jackson Hospital & Clinic in Central Alabama may initially use the RoM to enhance two particularly crucial quality of care areas: patient safety and satisfaction. Striving for greater objectivity and precision, they may wish to move away from their current qualitative review of "problem lists" through "likert scales" of 1 (least satisfied) to 5 (most satisfied), and various satisfaction tests to a more objective and automated approach that will identify threat zones, appropriate countermeasures and budgetary allocation for these countermeasures. These two hospital systems will work with the authors (Dr. M. Sahinoglu and Dr. Ken Wool) as a multidisciplinary team to implement the RoM and make the study medically viable. By assuring a high quality patient-centered level of care, they can improve patient care and hospital management. They further believe that these efforts will facilitate an environment of continuous improvement. From various meetings two recommendatory points were acknowledged. They will be addressed in stages 1 and 2 as follow:

Stage 1. The questionnaire should be transparent, objective and understandable to the full extent by the anonymous visitors of all levels of education in order to have a standard response base.

Stage 2. Costs associated with the threat factors that form the overall lump sum remediation cost to mitigate risk should be addressed by each hospital differently in the light of their operational conditions subject to a specific economic task-force analysis. The plans were mutually made to provide information on how to obtain the value of risk probabilities and redemption costs:

(a) An overall allocated (subject to feedback) lump sum cost to meet the countermeasures will be distributed regarding the entirety of threats for the vulnerabilities on an individual basis. This will not be made public to the patients taking the quality questionnaire but will be kept internally until the second stage risk mitigation procedures take effect by the hospital authorities.

(b) In estimating costs for threats, a task force in a particular hospital in conjunction with the hospital's IT personnel will analyze past and present costs adjusted by the inflation and depreciation factors to achieve this hurdle to minimize error. Since this is also a risk comparison effectiveness solution, the improvement can be cited as a percentage even if the dollar values are not exact.

It is planned that each participating hospital will create a web portal where individual patients can participate in the enhancement of the quality of patient-centered care by answering the assessment questions that provide the RoM with its input risk data once a large number of random sample size is attained. Although a random sample size of 15–30 is good enough to run statistical inferences utilizing Central Limit Theorem distributions, the proposed study wishes to have multiple samples of $15 < n < 30$ to have statistically robust estimators. The patient data will be amassed and an aggregate risk level determined. Implementing the RoM, they can then optimize the results, which present them with an objective (as opposed to subjective where human emotions are involved), econometric guide as to what countermeasures to apply to meet the identified threats and what funding to allocate for these countermeasures in which priority order. Additionally, hospitals can repeat the process on a periodic basis. Thus, with a baseline established and periodic assessments made, the hospitals can use this mechanism for continuous improvement, seeing where they rate currently versus previous time periods.

Conclusions

Can technology cure healthcare [32]? Medicine has been considered an *Art* for centuries and is finally moving into the molecular and microchip age. Likewise management of the business of medical care delivery is poised to make a quantum leap from the days of subjective decision making (educated guess work) to a new management paradigm of objective real time computer-generated risk and financially-based data [33–35]. This fact is evidenced by the practical applications of automated software such as what in the case of ROM can provide as hereby proposed [3], among others [36].

Acknowledgments I wish to acknowledge and thank Carole Wright at Wright & Associates (caw222@gmail.com) for her contributions in editing this applied research paper. Also Scott Morton at AUM's Informatics Institute (Smorton1@aum.edu) for his contributions in risk related topics is acknowledged with thanks.

References

1. National Survey on Consumers' Experiences with Patient Safety and Quality Information, Kaiser Family Foundation, Agency for Healthcare Research and Quality, November 2004.
2. M. Sahinoglu, "Security Meter- A Practical Decision Tree Model to Quantify Risk," *IEEE Security and Privacy*, 3 (3), April/May 2005, pp. 18–24.
3. M. Sahinoglu, Trustworthy Computing: Analytical and Quantitative Engineering Evaluation (CD ROM included), J. Wiley & Sons Inc., Hoboken NJ, www.johnwiley.com, August 2007.
4. P. Aspden, J.M. Corrigan, J. Wolcott, S.M. Erickson, editors. Institute of Medicine, Committee on Data Standards for Patient Safety. Patient safety: achieving a new standard of care. Washington, DC: National Academies Press; 2004.
5. National Healthcare Quality Report, U.S. Department of Health and Human Services, Agency for Healthcare Research and Quality, AHRQ Publication No.08-0040, February 2008.
6. L.T. Kohn, J. M. Corrigan, M.S. Donaldson, editors. Institute of Medicine, Committee on Quality of Health Care in America. To err is human: building a safer health system. Washington, DC: National Academies Press; 1999.
7. C. Zhan, M.R. Miller. Excess length of stay, charges, and mortality attributable to medical injuries during hospitalization. JAMA.; 290(14):1868–74, 2003.
8. R. I. Iyengar, and Y. A. Ozcan, "Performance Evaluation of Ambulatory surgery Centers: An Efficiency Approach", Health Services Management Research, 22(4):184–190, 2009.
9. "Hospitals Overhaul ERs to Reduce Mistakes", Wall Street Journal—Marketplace, May 10, 2011.
10. Dr. Paulo Franca, CIO-Data Care, Personal Communication, www.datacare.com, December 2010.
11. M. Sahinoglu, "An Input-Output Measurable Design for the Security Meter Model to Quantify and Manage Software Security Risk", *IEEE Trans on Instrumentation and Measurement*, 57(6), pp. 1251–1260, June 2008,.
12. Atul Gawande, "The Checklist Manifesto: How to Get Things Right (Hardcover)", Metropolitan Books, First Edition, 2009.
13. C. Pope and N. Mays, "Reaching the Parts Other Methods Cannot Reach: An Introduction to Qualitative Methods in Health and Health Services Research", Department of Epidemiology

and Public Health, University of Leicester, UK, BMJ (British Medical Journal) Vol. 311, 1, pp. 42–45, July 1995.
14. Yasar A. Ozcan, Quantitative Methods in Health Care Management: Techniques and Applications, 2nd Ed. Jossey-Bass/Wiley, San Francisco, CA. 2009.
15. Yasar A. Ozcan, Health Care Benchmarking and Performance Evaluation—An Assessment using Data Envelopment Analysis. Springer, Newton, MA, 2008.
16. Elena Andresen, Erin DeFries Bouldin, Public Health Foundations, Concepts and Practices, John Wiley and Sons, Inc., San Francisko, www.josseybass.com, 2010.
17. Lisa Wentzel, "Quantitative Risk Assessment", *Working paper for DISS 765 (Managing Risk in Secure Systems)*, Spring Cluster 2008 for Dr. Cannady, Nova Southeastern University, May 28, 2006.
18. E. Forni, *Certification and Accreditation,* AUM Lecture Notes, DSD (Data Systems Design) Labs, 2002.
19. Capabilities-Based Attack Tree Analysis, Retrieved from www.amenaza.com and http://www.attacktrees.com/, 2005.
20. D. Gollman, *Computer Security*, 2nd Ed., J. Wiley & Sons Inc., England, 2006.
21. M. Sahinoglu, "Security Meter Model- A Simple Probabilistic Model to Quantify Risk," *55th Session of the International Statistical Institute*, Sydney, Australia, Conference Abstract Book, p. 163, 2005.
22. M. Sahinoglu, "Quantitative Risk Assessment for Software Maintenance with Bayesian Principles," *International Conference on Software Maintenance*, Budapest, Hungary, ICSM Proc. II, 67–70, 2005.
23. M. Sahinoglu, "Quantitative Risk Assessment for Dependent Vulnerabilities," *The International Symposium on Product Quality and Reliability (52nd Year)*, Proc. RAMS06, New Port Beach, CA, 2006.
24. A. Arora, D. Hall, C. A. Pinto, D. Ramsey and R. Telang, "Measuring the Risk-Based Value of IT Security Solutions", *IT Prof.*, 6(6), 35–42, 2004.
25. D. Geer, Jr., Kevin Soo Hoo and Andrew Jaquith, "Information Security: Why the Future Belongs to the Quants", *Secure Computing Practices: Key Articles from IEEE Security and Privacy*, IEEE Computer Society, 2004.
26. D. Landoll, *The Security Risk Assessment Handbook*, Auerbach Publications, Taylor and Francis Group, Boca Raton, 2006.
27. S.C. Palvia, S. R. Gordon, "Tables, Trees and Formulas in Decision Analysis*, Commun. ACM*, 35(10), 104–113, 1992.
28. George Cybenko, "Why Johnny Can't Evaluate Security Risk", *IEEE Security and Privacy*, 4 (5), 2006.
29. Pete Lindstrom, "Yes, You Can Quantify Risk*", For Pete's Sake - ISSA Journal (Information Systems Security Association)*, www.issa.org, p. 9, April 2008.
30. "Wellpoint Shakes Up Hospital Payments", Wall Street Journal—Marketplace, May 16, 2011.
31. M. Sahinoglu, L. Cueva-Parra, D Ang (2012), "Game-theoretic computing in risk analysis", WIREs Comput Stat 2012, doi: 10.1002/wics, 1205.
32. Jacob Goldstein, "Can Technology Cure Health Care? How Hospitals Can Make Sure Digital Records Live Up to Their Promise, Because So far They Haven't", Wall Street Journal Report on Innovations in Health-Care, R3, 2010.
33. D. W. Bates, "Physicians and Ambulatory Electronic Health Records", 2005, Health Affairs (24)5: 1180–1189.
34. B.L. Chang, S. Bakken, S. S. Brown, et al., "Bridging the Digital Divide: Reaching Vulnerable Populations", 2004, J AM Med Inform Assoc. 11:448–457.
35. Institute of Medicine, "Priority Areas for National Action: Transforming Healthcare Quality", Washington DC, National Academy Press, 2003.
36. Irem Ozkarahan, "Allocation of Surgeries to Operating Rooms Using Goal Programming", Journal of Medical Systems, Vol. 24, No. 6, pp. 339–378, 2000.

Chapter 14
Use of Session Initiation Protocol in Multimedia Communications: Evaluation of Server Performance Based on Software Profiling

Mansour H. Assaf, Maquada Williams, Sunil R. Das, Satyendra N. Biswas and Scott Morton

Introduction

The session initiation protocol (SIP) is the accepted standard for multimedia conferencing using internet protocol (IP) proposed by the Internet Engineering Task Force (IETF) responsible for developing and promoting internet standards. The SIP is an American standard code for information interchange (ASCII)-based, application layer control protocol in the seven-layer open system interconnect (OSI) model, defined in request for comments (RFC) 2543 that can be used for application layer signaling and describes initiation, modification and termination of interactive multimedia communication sessions between the users [1–18]. Designed to be independent of the underlying transport layer, it can run on transmission control protocol (TCP) or user datagram protocol (UDP). The SIP is primarily designed for setting up voice or video calls and is used where session initiation is a requirement in an application. These applications include event

M. H. Assaf
Department of Electrical and Electronic Engineering, University of the South Pacific, Suva, Fiji

M. Williams
Department of Information and Communication Technology, University of Trinidad and Tobago, Trinidad, West Indies

S. R. Das (✉) · S. Morton
Department of Computer Science, College of Arts and Sciences, Troy University, Montgomery, AL 36103, USA
e-mail: sdas@troy.edu

S. R. Das
School of Information Technology and Engineering, Faculty of Engineering, University of Ottawa, Ottawa, ON K1N 6N5, Canada

S. N. Biswas
School of Engineering and Computer Science, Independent University, Dhaka 1329, Bangladesh

S. C. Suh et al. (eds.), *Applied Cyber-Physical Systems*,
DOI: 10.1007/978-1-4614-7336-7_14,
© Springer Science+Business Media New York 2014

169

subscription and notification, terminal mobility and so on. There exist large numbers of SIP-related RFCs that define behavior for such applications.

All voice and video communications are done over separate session protocols [2]. The addressing of the functions of signaling and session management by the SIP and other voice over internet protocols (VoIPs) takes place within a packet telephony network. The application layer signaling allows call information to be carried across the network boundaries. The session management, viz. modification and termination of interactive multimedia communication sessions between users, as the name suggests, simply provide the ability to control the attributes of an end-to-end call [6].

The SIP provides capabilities for:

- Determining the location of the target end point with support address resolution, name mapping and call redirection.
- Ascertaining the media capabilities of the target end point via session description protocol (SDP); it locates the *lowest level* of common services between the end points. The conferencing is established using only the media capabilities that can be supported by all of the end points.
- Deciding upon the availability of the target end point. If a call cannot be completed because the target end point is unavailable, the SIP determines whether the called party is already on the phone or did not answer in the allotted number of rings. It then returns a message indicating why the target end point was unavailable.
- Establishing a session between the originating and target end points. If the call can be completed, the SIP establishes a session between the end points. It also supports mid-call changes, such as the addition of another end point to the conference or changing of a media characteristic or coder-decoder (codec).
- Handling the transfer and termination of calls. The SIP supports the transfer of calls from one end point to another. During a call transfer, the SIP simply establishes a session between the transferee and a new end point (specified by the transferring party) and terminates the session between the transferee and transferring party. At the end of a call, the SIP terminates the session among all the parties.

A conferencing can consist of two or more users and can be established using multicast or multiple unicast sessions [6]. In the present paper, we investigate the enhancements made to the SIP and multimedia conferencing.

SIP Protocol: A Brief Overview

The SIP is an ASCII-based protocol, as mentioned, that uses requests and responses to establish communication among components in the network and/or to set up a conference between two or more end points [6]. The protocol is designed to focus on call setup and signaling. However, with the addition of network elements like the proxy servers and user agents, the SIP can now afford or support call processing functions and other features. Though, the SIP is a peer-to-peer protocol, viz. one in which two end points can communicate without the intervention of the SIP architecture, many advanced call processing features can now be implemented in an SIP-enabled telephony. This action requires only a simple core network with intelligence distributed to the network cycle embedded at the end points [2]. The users in an SIP network are identified by unique SIP addresses. An SIP address is similar to an e-mail address and is in the format of sip:userID@gateway.com [6].

Establishing communications using the SIP usually occurs in six steps, as mentioned below [1]:

- Registering, initiating and locating the user.
- Determining the media to use—this involves delivering a description of the session that the user is invited to.
- Finding out the willingness of the called party to communicate—the called party must send a response message to indicate the willingness to communicate— accept or reject.
- Call modification or handling—for example, call transfer (optional) or call termination.

The users register with a register server using their assigned SIP addresses. The register server provides this information to the location server upon request. The registration takes place each time a user turns on the user-client application or when the SIP user needs to inform the server of its location.

When a user initiates a call, an SIP request is sent to an SIP server (either a proxy or redirect server). The request includes the addresses of the caller and of the intended callee. Over time, an SIP end user might move between the end systems. The location of the end user can be dynamically registered with the SIP server. The location server can use one or more protocols, viz. finger and referral whois (rwhois) to locate the end user. Because the end user can be logged in at more than one station and the location server can sometimes have inaccurate information, it might return more than one address for the end user. If the request is coming through an SIP proxy server, the proxy server will try each of the returned addresses until it locates the end user. If the request is coming through an SIP redirect server, the redirect server forwards all the addresses to the caller in the contact header field of the invitation response [6].

Hardware Components in an SIP

Though a peer-to-peer protocol, the SIP requires network elements to work as a practical service and its distributed architecture includes components like the location server, redirect server, register server, user agent and proxy servers.

The peers in a session are called user agents (UAs). A user agent can function in one of the following roles:

- User agent client (UAC)—A client application that initiates the SIP request.
- User agent server (UAS)—A server application that contacts the user when an SIP request is received and then returns a response on behalf of the user [6].

The user agents initiate, receive and terminate calls [2]. Typically, an SIP end point is capable of functioning as both an UAC and UAS, but functions only as one or the other per transaction. Whether the end point functions as an UAC or UAS depends on the user agent (UA) that initiated the request [6].

From an architectural standpoint, the physical components of an SIP network can be grouped into two major categories: clients and servers. Figure 14.1 illustrates the architecture of an SIP network.

The SIP clients include:

- Phones—which can act as either the UAS or UAC. The softphones {personal computers (PCs) that have phone capabilities installed} and Cisco SIP IP phones can initiate the SIP requests and respond to the requests [6].
Gateways—these can provide call control. The gateways incorporate many services, the most common being a translation function between the SIP conferencing end points and other terminal types. This function includes translation between transmission formats and communication procedures. In addition, the gateways translate between the audio and video codecs and perform call setup and clearing at both the local area network (LAN) side and switched-circuit network side [6].

The SIP servers include the following:

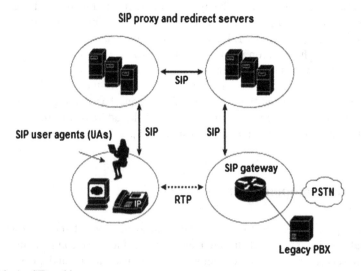

Fig. 14.1 An SIP architecture

- Proxy server—it is an intermediate device that receives the SIP requests from a client and then forwards the requests on client's behalf. Basically, proxy servers receive the SIP messages and forward them to the next SIP server in the network [6]. The proxy servers are programs that act as servers and/or clients making requests on behalf of other clients [2]. The proxy servers also help route requests, authenticate users for servers, implement call routing policies and provide features to users [1].
- Location servers—these are used to obtain information about a called party's possible location [1].
- Redirect server—it provides the client with information about the next hop or hops that a message should take; then, the client contacts the next hop server or UAS directly [6]. The redirect servers accept SIP requests, can map the addresses into new addresses and return the new addresses to the client. These servers do not initiate an SIP request and can neither terminate nor accept calls [2].
- Register server—this server processes requests from UACs for registration of their current location. The register servers are often co-located with a redirect or proxy server [6].

Using a Proxy Server

If a proxy server is used, the caller UA sends an INVITE request to the proxy server; the proxy server determines the path and then forwards the request to the callee. The callee responds to the proxy server, which, in turn, forwards the response to the caller. The proxy server forwards the acknowledgments of both parties. A session is then established between the caller and callee. The real-time transfer protocol (RTP) is used for the communication between the caller and callee, as shown in Fig. 14.2.

Using a Redirect Server

If a redirect server is used, the caller UA sends an INVITE request to the redirect server; the redirect server contacts the location server to find the path to the callee and then, the redirect server sends that information back to the caller. The caller then acknowledges receipt of the information. The caller next sends a request to the device indicated in the redirection information, viz. the callee or another server that will forward the request. Once the request reaches the callee, it sends back a response and the caller acknowledges the response. The RTP is used for communication between the caller and callee. Figure 14.3 shows an SIP session through a redirect server.

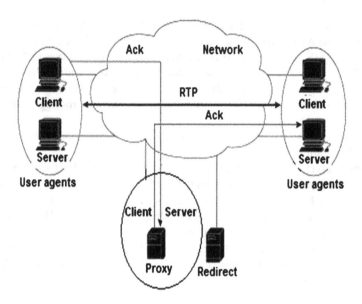

Fig. 14.2 An SIP session through a proxy server

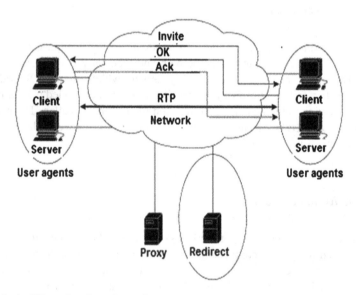

Fig. 14.3 An SIP session through a redirect server

VoIPs allow users to make phone calls over the internet using the packet-switched network as a transmission medium. The maturity of VoIP standards such as SIP [8] and quality of service (QoS) on IP networks ensure:

- maximizing network efficiency;
- streamlining network architecture;
- reducing capital costs;
- minimizing operational costs; and
- opening up new service opportunities such as multimedia conferencing.

QoS is considered to be one of the key features of voice and video over IP communications. Unfortunately, the active control of network resources within the IP transport network [9] as performed according to next generation network (NGN) QoS architecture [10] results in a considerable amount of resource management traffic which is not scalable with the increasing number of NGN subscribers. In order to address the issue, an integrated framework for comprehensive QoS control in SIP-based NGN has been introduced in [11]. A characteristic of this framework is the continuous collection of information on IP transport network performance as experienced by any NGN subscriber terminal. The network performance of an IP transport network is characterized by packet loss ratio, transfer delay and delay variation [12, 13]. These network performance parameters substantially influence the QoS of a real-time communication service as experienced by its users.

In the application area of VoIP and SIP, traffic behavior and failures in particular software implementations were investigated. The general profiling methodology has been applied to analyze SIP protocol to illustrate the characteristics of SIP traffic in a real VoIP network and to use them in justifying the statistics and features we have taken for profiling SIP traffic behavior. In normal operational environments, SIP traffic behavior tends to be very stable in terms of various SIP message types, user registrations, calls and other related activities [14]. In [15], the authors describe their solution for profiling SIP-based VoIP traffic to automatically detect anomalies. They demonstrate that SIP traffic can be modeled well with their profiling and that anomalies could be detected. In [16], formal testing of an open source and a commercial SIP proxy leads to errors with SIP registrar. Both findings are encouraging to propose a method for not only detecting incompatibilities and testing SIP proxies but also to providing solutions for messages rejected due to slightly different interpretations of standards or software faults. In [17], a *fuzzer* was added to an SIP proxy installation. The system was used to systematically test an SIP proxy with different faulty messages in terms of syntax and protocol behavior. In [18], however, both incoming and outgoing SIP messages of a proxy were analyzed by an in-kernel linux classification engine. In the present study, on the other hand, a rule-based approach has been used, the rules being pre-defined, as will be evident.

Software Profiling

The performance analysis or *profiling* is a form of dynamic program analysis where the program's behavior is analyzed using information gathered as the program runs. The goal of performance analysis is to determine which parts of a program need to be optimized for speed and memory usage. The performance analysis tools measure the frequency and duration of the function calls as the program runs. The data are collected using hardware interrupts, code instrumentation, operating system hooks and performance counters with output of a trace or projector. This process is called the performance engineering process.

Three methods of gathering the data exist, viz. event-based profiles, statistical profiles and instrumentation [3].

SIP Application Server Performance and Benchmarking

An SIP is defined or observed in the context of VoIP instant messaging, video conferencing and others. The Java community has created and standardized an application programming interface (API) that facilitates building the SIP applications with Java 2 platform enterprise edition (J2EE) application servers. The performance engineering of the SIP applications in a J2EE environment is discussed next.

An SIP is noted for its decentralized approach to session negotiation and moves control handshaking to end points rather than centralized control, rendering it extensible, scalable and useful for mobile applications. The use of an SIP in applications that interact using the SIP and hypertext transfer protocol (HTTP) is also highlighted. The differences between the two protocols (SIP and HTTP) are discussed in an effort to guide through the performance engineering and are listed below [5]:

- The SIP has quality of service (QoS) latency requirements on protocol responses.
- The SIP has a peer-to-peer architecture, while an HTTP utilizes a client–server approach.
- The SIP servlet may also act as an SIP client.
- The transactions as well as sub-transactions may be stateless or state-full.
- The interactions with an HTTP are only synchronous whereas with an SIP, they may be either synchronous or asynchronous.
- More than one response may be generated to a single request and a single SIP request may be served by many SIP applications in the same application server.

SIP Performance Input Factors

There are several factors that contribute to the performance and describe the overall behavior of the SIP application. An application latency budget is the mean delay with a standard deviation as well as percentile delay of 99.999 through the application, with the main factor in this latency being the garbage collection time in a virtual machine-based application.

The system garbage collection latency budget is typically more flexible than the application latency budget because there is some elasticity in the system.

The system configuration refers to the workload or whether or not a high availability configuration is needed to avoid a single point of failure that would determine if a cluster of servers or a single server is used.

The types of application—there are several canonical architectures for the SIP applications and each has its own performance profile.

The maximum SIP message rate—this is the rate at which the SIP messages are presented to the application. The application's processing rate must exceed this value.

The maximum session creation rate—it governs the rate of memory consumption and central processing unit (CPU) utilization due to the memory management activity required for the new session objects.

The duration the session lasts—the user sessions can expire due to a default session time-out value or when the end points terminate the session. The session duration, session creation rate and per session memory consumption drive working memory requirements for all system components, which maintain session state.

The maximum application session memory consumption—it represents the amount of memory consumed when a user session is created, added to the memory used during the lifetime of the user session (Kbytes/application).

The average CPU utilization—this specifies a budget for the CPU consumption, typically less than 100 %, to allow for the management and operating system activity.

The transport type—the UDP is the primary transport protocol used in the SIP with a TCP, if the SIP messages are too large or if data encryption is needed for security.

The authorization—it is performed by the SIP application and adds to the performance overhead of the system.

SIP Performance Environment and Tuning Factors

An execution environment is required for the deployment and tuning of an SIP application. One important element in the performance environment is the number of servers used, which can be determined by the SIP message rate or maximum session creation rate. Another factor of importance is the number of CPUs per server, as more processors translate to faster processing of the SIP stack and other maintenance activities. The number of the SIP application processes per server and the virtual machine tuning values are also important to the performance environment and tuning factors.

Many tuning options that affect latency message throughput and memory consumption revolve around reducing the garbage collection and latency. The performance areas that are important for an SIP are as follows: the heap size or working memory for creating the SIP application objects. The heap size directs the number of concurrent SIP sessions and garbage collection. It is important to select an algorithm with the lowest latency and/or most deterministic latency during the application. A garbage collection thread will take control of all available CPUs for the purpose of performing in the shortest time possible. Hence, the importance of the number of concurrent, foreground threads cannot be ignored. The number of concurrent background garbage collection threads is also important as idle CPU periods during the execution of the application, where the background garbage collection threads can perform preparatory work.

SIP Performance Engineering Process

The engineering process is a methodical approach that manages all the factors that affect the performance. The methodology for analyzing the SIP performance is comprised of:

- Pre-screening the input data to determine if the measurements are valid.
- Processing the input data for a single run, either as a time series graph or in terms of statistics generated for various factors during the steady-state portion of the measurements.

SIP Stone: Benchmarking SIP Server Performance

A set of metrics for evaluating and benchmarking the performance of the SIP proxy, redirect and register servers is used in the paper. The benchmark is limited to evaluating the sustainable rate of a representative workload with registrations, successful forwarding and unsuccessful call attempts. The benchmark is an attempt at characterizing the server performance in a way that is useful for dimensioning and provisioning. The self-imposed limitations of the benchmark include the fact that the benchmark does not evaluate protocol compliance or robustness and is only one metric out of many in evaluating the servers and does not reflect functionality.

Issues Involved in Designing SIP Benchmarks

The benchmarks are designed to help understand how a telephony system will perform and to help in the comparison of different types of implementations. The primary metric in performance engineering systems is the number of requests processed successfully within a given delay constraint. The performance of a server depends on the implementation, size of the user population serviced, number of domain name system (DNS) lookups required, transport protocol use and type and statistical arrival process of requests submitted.

The issues in designing a benchmark include repeatability, viz. the ability to replicate the results by a third party when the testing is done under a specified environment and on the measurability which defines the ease with which the results can be quantified to present a picture of the system being evaluated. The scalability is another issue that should be considered in designing a benchmark; this refers to the benchmark's ability to run on different configurations.

The benchmarking environment consists of test runs with load levels generated by load generators and targeted at the SIP server being tested, the state in which the UACs generate the SIP request and/or state during the execution of the benchmark from which the reported performance metric is derived. The transaction response time is the elapsed time between the first byte sent by a UAC to initiate a transaction to the last byte received by the UAC that ends the transaction. The number of registrations per second defines the average number of successful registrations per second during the measurement interval. The number of calls per second is the mean number of calls per second completed with a 2xx or 4xx response during a measurement interval. The transaction failure probability is defined as the fraction of transactions that fail.

Measurement Methodology

The registrations per second (RPS) and calls per second (CPS) values are determined by increasing the request rate until the transaction failure probability (TFP) has increased by 5 % during an interval of 15 min. The benchmark would be the highest sustained throughput reported. The scaling requirements are also considered in designing a benchmark for the SIP server performance. The size of the user population used for the tests should scale with the request handling capacity of the server. The response time constraints are also important; therefore, during each measurement interval, at least 95 % of call setup requests of each type must have a transaction response time (TRT) of less than the specified constraint.

The metrics and parameters to be reported on are:

- Description of the clustering configuration including the number of hosts and load balancing mechanism used.
- All aspects of server hardware (CPU count, type, speed, memory configuration, etc.).
- The server operating system, version and any non-standard tunings or settings.
- The server software configuration.
- The number and type of the SIP stone load generators that participate in the run and number of worker threads created in each generator.
- The number and type of network segments connecting the load generators to the SIP server.
- The arrangement of load generators on the network segments.
- The workload parameters that may be increased across several test runs.

The benchmark results to be reported should include:

- The number of connections requested by the clients and accepted by the SUT per second.
- The CPU and memory utilization of the server at various loads.
- Curve plotting the token rotation time (TRT) as a function of the request arrival rates, CPS and RPS.

Experimental Results

We used the SIP source code version that was used in the implementation of yet another telephony engine (YATE) [4]. The YATE is a next-generation telephony engine that is currently focused on VoIP and public switched telephone network (PSTN). The YATE can be used as a VoIP server, VoIP client, VoIP to PSTN gateway, SIP proxy, SIP router and SIP registration server.

The genuinely not Unix (GNU) gprof [7] was used as a profiling tool in evaluating the YATE source code performance. The flat profile shows the total amount of time a given program spends executing each function separately. The call graph shows how much time was spent in each function and its children.

The profiling works by changing how every function in the program is compiled so that when it is called, it will stash away some information about where it was called from. The profiling also involves watching the program as it runs and keeping a histogram of where the program counter happens to be every now and then [7].

For the purpose of this work, the SIP profile data were collected and analyzed for three different scenarios:

SIP scenario 1: Callgrind.out.8317—the profiler was run for 5.12 min.
SIP scenario 2: Callgrind.out.8916—the profiler was run for 41.14 s.
SIP scenario 3: Callgrind.out.8967—the profiler was run for 5.12 min.

A summary of obtained sample data outputs is shown in Tables 14.1, 14.2 and 14.3.

Figures 14.4, 14.5 and 14.6 break down these results into three categories:

- Flat profile analysis.
- Callees—comparison of TelEngine::Mutex::Mutex()<cycle1>.
- Callees—comparison of TelEngine::Mutex::Mutex(bool)<cycle1>.

The flat profile analysis shows how the SUT in general and SIP server in particular behave under various loads and resource constraints. Two callee functions (TelEngine) are also evaluated under different loads; the number of instruction read (IR) and number of callees (count) are recorded and compared. The performance measure shows that the SIP passes specified server benchmark test.

Table 14.1 Data output in the flat profile for Callgrind.out.8317

0.18 units of time were spent in

TelEngine::SIPEngine::getEvent() (cycle1) and all its children

0.18 units were also spent in that function without its children

Time spent in other five functions is negligible

SIPEngine and ~SIPEngine functions were both called once

Callees

TelEngine::Mutex::Mutex()<cycle1>, TelEngine::Mutex::Mutex(bool)<cycle1> were also
    called once

All Callees

Various amounts of time were spent in each of the following functions called once: snprint,
    vtprint, rand, random, _IO_setb, TelEngine::Objlist::Objlist(),
    TelEngine::DebugEnabler::debuglevel(int)

_IO_default_xsputn, strchmul and TelEngine::String::String() were called 3, 2 and 3 times,
    respectively, and 2 unnamed functions were called 16 and 4 times, respectively. 285.71 units
    were spent in all called functions, with the most time being spent in an SIP related function

Table 14.2 Data output in the flat profile for Callgrind.out.8916

The values were the same except for the first row

TelEngine::SIPEngine::getEvent()<cycle1>were called only 13711 times less than before

Callees

Contained the same functions as in the previous profile data outputs. All functions were called
    13711 times which is 1241 less than 8317

All Callees

Showed on unnamed functions their Incl and Self values 3.92, the same as that for 8317

Table 14.3 Data output in the flat profile for Callgrind.out.8967

The values were almost the same as for 8316 and 8317

Here, Incl and Self values for TelEngine::SIPEngine::getEvent()<cycle1> were 0.17 and 0.16,
    respectively

This function was called 18641 times

All callees

Showed on unnamed functions their Incl and Self values as 3.92, the same as that for 8317

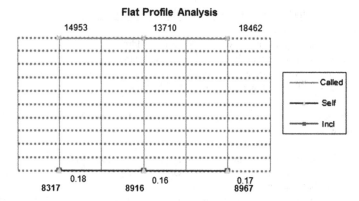

Fig. 14.4 Flat profile analysis

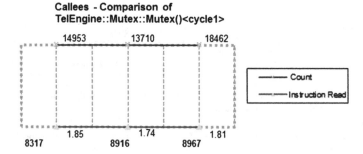

Fig. 14.5 TelEngine::Mutex::Mutex()<cycle1>

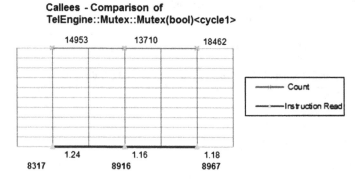

Fig. 14.6 TelEngine::Mutex::Mutex(bool)<cycle1>

Conclusions

The chapter subject various server test scenarios to evaluate the efficiency and behaviors of the SIP. The study utilizes benchmark tests to analyze the performance of the SUT. The numerical computations carried out illustrate the impact of large numbers of calls and explore the effects of loads on the SIP by measuring individual function's run time together with total time to execute the software program.

The chapter investigates the SIP servers and enhancements made to internet telephony protocol, which is specially designed for multimedia conferencing. Based on the study of the SIP servers, it was noted that the complexity, computation, memory capacity and CPU speed associated with the hardware architecture pose challenges, particularly from the viewpoint of resource utilization, dynamic adaptation capabilities and real-time implementations. However, the SIP protocol did pass the benchmark test according to the data collected from the flat profile analysis.

Acknowledgments This research was supported in part by the Department of Computer Science, College of Arts and Sciences, Troy University, Montgomery, AL 36103 USA.

References

1. Voice over IP Protocols–An Overview–www.vovida.org.
2. Session Initiation Protocol–http://en.wikipedia.org/wiki/Session_Initiation_Protocol.
3. Profiling (Computer Programming)–http://en.wikipedia.org/wiki/Performance_analysis.
4. Yet Another Telephony Engine–http://yate.null.ro/.
5. A Tutorial on SIP Application Server Performance and Benchmarking–www.cs. manchester.ac.uk.
6. Overview of the Session Initiation Protocol–http://www.cisco.com/univercd/cc/td/doc/ product/voice/sipsols/biggulp/bgsipov.htm.
7. GNU gprof–http://www.cs.utah.edu/dept/old/texinfo/as/gprof.html.
8. J. Rosenberg, H. Schulzrinne, G. Camarillo, A. R. Johnston, J. Peterson, R. Sparks, M. Handley, and E. Schooler, SIP: Session Initiation Protocol (RFC 3261)–Internet Engineering Task Force, 2002.
9. F. Weber, W. Fuhrmann, W. Trick, U. Bleimann, and B. Ghita, "QoS in SIP-based NGN– State of the art and new requirements", Proceedings of the 3rd Collaborative Research Symposium on Security, E-learning, Internet and Networking, 2007, pp. 210–214.
10. ETSI TS 185 001 V1.1.1, Technical Specification, "NGN quality of service (QoS) framework and requirements", 2005, ETSI TISPAN.
11. F. Weber, W. Fuhrmann, U. Trick, U. Bleimann, and B. V. Ghita, "A framework for improved QoS evaluation and control in SIP-based NGN", Proceedings of the 7th International Network Conference, 2008, pp. 27–37.
12. A. Kos, B. Klepec, and S. Tomazic, "Techniques for performance improvement of VoIP applications", Proceedings of the 11th Mediterranean Electrotechnical Conference, 2002, pp. 250–254.

13. C. M. White, J. Raymond, and K. A. Teague, "A real-time network simulation application for multimedia over IP", Conference Record of the 38th Asilomar Conference on Signals, Systems and Computers, Volume 2, 2004, pp. 2245–2249.
14. H. J. Kang, Z.-L. Zhang, S. Ranjan, and A. Nucci, "SIP-based VoIP traffic behavior profiling and its applications", NARUS Technical Report, 2006, pp. 1–11.
15. H. Kang, Z. Zhang, S. Ranjan, and A. Nucci, "SIP-based VoIP traffic behavior profiling and its applications", MINENET, 2007, pp. 39–44.
16. B. Aichernig, B. Peischl, M. Weiglhofer, and F. Wotawa, "Protocol conformance testing a SIP registrar: An industrial application of formal methods", 5th IEEE International Conference on Software Engineering and Formal Methods, 2007, pp. 215–224.
17. H. Abdelnur, R. State, and O. Festor, "KiF: A stateful SIP fuzzer", Proceedings of the 1st International Conference on Principles, Systems and Applications of IP Telecommunications, 2007, pp. 47–56.
18. A. Acharya, X. Wand, C. Wright, N. Banerjee, and B. Sengupta, "Real time monitoring of SIP infrastructure using message classification", MINENET, 2007, pp. 45–50.

Chapter 15
Principle of Active Condition Control: Impact Analysis

Igor Schagaev and John N. Carbone

The Cycles of Maintenance

Existing regulations require that maintenance for aircraft should be performed periodically according to the schedule defined using manufacturer data. Maintenance periods are accompanied by intermediate checks based on the actual load and annual checks [1–4]. Unfortunately, as outlined by [5–8], only a small proportion of world aircraft fleet are maintained according to this schedule.

The lack of an effective policing of maintenance and safety requirements in aviation is a major contributory factor for poor safety and thus provides little benefit for aviation [9, 10]. When safety checks are mandatory and performed by an independent body a certificate for permitted vehicle use is issued. Regretfully, the coverage of checking is highly unlikely to be considered as complete [11], making risk of aircraft use substantial and unavoidable.

Even properly maintained aircraft *on the ground* does not guarantee reliability and safety of an aircraft *during flight*. Until now neither control nor flight safety management system has taken into account an information about faults that the aircraft may already have; does not prove or monitor quality of maintenance, does use in real time structural models of aircraft and does check deviations that are developing. This creates a situation where the decision to use the aircraft for the next flight is taken almost voluntarily, based more or less on trust. Note that the quality of certification depends heavily on human factors (existing qualification, training, integrity etc.). The "Observer" publication (21st Aug 2005: "Airline pilots 'lack basic skills'") revealed that the risks associated with poor training are real concern in the CA segment). In turn, recent accidents: June 2009 (A330 AirFrance), November 2010

I. Schagaev (✉)
Faculty of Computing, London Metropolitan University, London N7 8DB, UK
e-mail: i.schagaev@londonmet.ac.uk; info@it-acs.co.uk

J. N. Carbone
Raytheon Tactical Intelligence Systems Garland, Garland, TX 75040, USA
e-mail: jcarbone@raytheon.com

S. C. Suh et al. (eds.), *Applied Cyber-Physical Systems*,
DOI: 10.1007/978-1-4614-7336-7_15,
© Springer Science+Business Media New York 2014

A380, Boeing 747 (Quantas), 2012 complete mishap with A380 wings show that neither design of aircraft nor their control systems are satisfactory reliable.

Two idealistic approaches that might improve maintenance and aviation safety have been pursued so far: (a) changing human nature by special training and retraining (i.e. unfounded optimism) or, (b) changing the world (i.e. improving the quality of maintenance and upgrading landing strips to airfields with proper maintenance facilities), making maintenance obligatory—neither is realistic nor feasible.

What *is* possible? An answer is in a designing a CPS system that is able to perform high quality analysis of aircraft conditions using accumulated and current flight (or mission) data from aircraft devices and knowledge of aircraft structure. Existing and new information technologies might be extremely helpful to implement this goal by making device and software for this kind of monitor. The results of this real time monitoring of conditions, when necessary, could supply relevant information about the current state of an aircraft for flight crew on board and operators, maintenance team, insurers and designers on ground. This allows correct decisions and "prescribing" procedures for aircraft maintenance. Above all, this analysis can run continuously on board and request recovery or servicing when necessary during and after flight.

The concept of preventive maintenance [10] has been known amongst aviation academics for a long time, but was never actually implemented [7]; two accidents with Rolls Royce engines with two days of 4th and 5th of November 2010 manifest the lack of knowledge and ability to apply them to keep required level of reliability for aircraft engines. To some extent preventive maintenance is progressing in the automotive sector, mostly for aggregation of information of wear of parts and the amount of vehicle use [6], but, again, volume of recalled cars due to poor reliability for Toyota, Mercedes and other brands exceeds hundreds of thousands every year, manifesting that existing concepts of preventive maintenance and quality of design are not sufficient or efficient.

The approach proposed here is called *principle of active condition control* (PACC), *concept of active system safety* was registered 20/09/2010 by European OHIM, No 008895674 and patented [14]. At the same time, no matter how good principle was introduced without implementation it has largely rhetoric value. To be implemented PACC must include model of aircraft feasible for real-time application, special on-board hardware and system software. This includes continuous, detailed dependency capture and analysis during development cycle, combined with PACC aircraft model, and combined with real-time analytic focused aggregation and processing of real-time aircraft data. Note that a pilot can't be involved in handling critical conditions—processes and complexity of control systems as well as aircraft designs do not leave a room for manoeuvre: humans become a weakest link and can't be considered as an element of active conditional control approach. This system has to monitor aircraft (or vehicle) conditions, call it *active condition control monitor* (ACCM). To have any credibility, ACCM itself must be ultra reliable in three ways:

1. Always be *available*, even though the aircraft itself may not be serviced to schedule.
2. Always offer *safe and relevant actionable advice* based on the current conditions, using previous flight data, current flight data and trustworthy analysis.
3. Present *an action plan to conserve or improve conditions by avoiding risk*, which is credible in its own right and transparent and clear to the operators, crew and other relevant institutions.

There are some challenges regarding determination of conditions of aircraft during flight: the amount of flight data available is approaching hundreds of megabytes, the complexity of fault free models of aircraft is growing, whilst while modelling of deterioration of aircraft conditions is an order of magnitude more complex.

But PACC has no palliatives: it only has abilities to determine a vehicle conditions and to react timely on their deterioration lowering the risk of use.

Secondly, the reliability of the existing parts of the aircraft will not be improved in the foreseeable future; in fact, they will gradually degrade due to aircraft aging and exploitation. In turn, complexity of modern aircraft complicates an overall reliability improvement.

Thirdly, the reliability of any safety and reliability control system must itself be extremely high ("who watches the watchers?") and faults possible in it should be isolated in terms of impact on aircraft operation. This kind of systems has to function over the whole life cycle of aircraft, without maintenance ("zero maintenance" approach was proposed by author of this paper in 2007 [15]).

So far, 'common sense' suggests an improvement of reliability and safety level using the aircraft's actual use and then advising on reliability and safety of its future use. This introduces the need of the continuous and instantaneous assessment of the aircraft reliability. Thus, to implement active conditional monitoring one to use current and accumulated flight data and create a model of aircraft, capable of assessing point availability in real time. Additionally, to produce a quality real-time result a CPS system framework must be instituted to reliably handle the vast information ingest and data interchange. Simultaneously, the framework must analytically process fast enough to provide a productive instantaneous assessment of the situation and thus an actionable predictive human usable result. Using this might improve *mission reliability*, i.e. the probability of successful completion of the flight. Above all, it is necessary to predict potential risks/faults and anticipate corrective or preventative action to improve/maintain safety of operation and its successful completion.

Information Content Management and Active Conditional Control

Modelling dependencies of vast arrays of components within an aircraft for PACC is arduous and complex. Here we discuss how PACC can be applied to existing designs and discuss the added benefits of planning for PACC from the beginning of a new design cycle to achieve optimal performance. Hence, how much knowledge is enough knowledge? What threshold of knowledge must be achieved about a system or a set of its components to make an informed PACC decision? How coupled or decoupled is the existing design? These are questions which have discrete answers when discussed within discrete contexts. For example, a system might have a functional requirement to include an oil pump. The pump will be rated as viable to a certain amount of use/miles/flight hours etc. and hence, due to the imperfect nature of reality, has a set of design parameters which provide information about a range of usage as opposed to an exact time. The oil pump is also a core sub-component of a larger system which has its own range(s) and set of independence and inter dependencies. Historically, Complexity Theory [16] provides solutions to minimizing information content and understanding design ranges, functional requirements, dependencies, design parameters, & constraints, as well as, the coupling and decoupling within a design. PACC takes advantage of complexity theory by maximizing effectiveness thru minimizing the amount of information content, as shown in Fig. 15.1, necessary to understand a situational range and to solve the right problem. If PACC planning is performed early during the beginning of a design cycle, an optimized model is produced a priori, and hence PACC has a more accurate model as initial input. This minimizes the time and analysis required to implement PACC for a given design.

Preventive Maintenance Versus Active Conditional Control

Current monitoring and maintenance systems do not provide in-depth knowledge of aircraft conditions; they suffer from latent (hidden) faults and therefore do not prevent or reduce the degradation of safety. In principle, any conditional monitoring system is implementing generalised algorithm of fault tolerance (GAFT) as introduced in [12], (see Fig. 15.1). In such systems, steps A, B, D, and E in Fig. 15.2 are not implemented in real time of mission. It is clear though that real time implementation of GAFT is essential for the purpose of active condition control. PACC implementation includes a use of several types [12] of redundancies

Fig. 15.1 Information axiom minimizing information content [17]

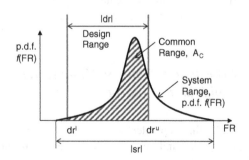

LOOP
A: Evaluate the conditions and processes in the system that create or might create a reduction of
        the current or future safety or other properties (diagnosis and prognosis).
B: Decide about trends in the system in terms of condition change (and level of danger/risk)
        using discrete, semantically driven or probabilistic models of the system (or combinations of
        them).
C: Determine of the reasons (or faults, or event) that cause a detectable reduction or
        deterioration of conditions or safety level.
D: Analyse the possible reactions and options available, including full or incomplete recovery
        (management of system deficiency).
E: Form the set of actions to restore and/or recover conditions (or safety).
F: Estimate of the level of safety achieved (restored and/or recovered).
END

Fig. 15.2 The algorithm to implement PACC

deliberately introduced in the system for implementation of steps of the algorithm Fig. 15.2. However, the choice of redundancy limits the design process when new features of an object are pursued.

The obvious question is: how active conditional control affects reliability and scheme of maintenance of an object? A simple answer is identification of condition or state and actions to tolerate/reduce consequences makes possible to avoid risky developments and, therefore, reduce harm and increase safety. An analysis of the potential for reliability gain from PACC implementation is the goal of this work.

The primary functions of ACCM are the evaluation of conditions and when necessary execution of preventive maintenance. Maintenance here is considered in a broad sense including PACC implementation of maintenance-on-demand during and after flight as well as an increase of quality of periodic maintenance.

An aircraft is an object, with cyclic operation that in principle includes preventive maintenance procedures. In practice it is hardly the case. The approach to periodic maintenance of aircraft is based on assumptions (which are sometimes quite naive and over-optimistic) about the guaranteed high quality of maintenance. Even when this periodic maintenance does take place the resulting state of an aircraft is very difficult to analyse. Additionally, flight information, estimation of condition of aircraft, its main structural elements as a system does not correspond to *before*, *during*, and *after* flights periods.

Preventive maintenance for aircraft, as well as for other complex technological objects with safety–critical functionality, was introduced in the early 1960s [10]. A simple Google search yields 1.3 million references for preventive maintenance. Aviation-related preventive maintenance is discussed at least 96 K references. At the same time, theory of preventive maintenance is mentioned in less than 100 references.

A possible reason for this gap difference is in the fuzziness of the meaning of "preventive maintenance" and the justification of its proper application. Usually those who use the term consider "preventive maintenance" from the position of business school courses for managers of airports and aircraft service centres.

The real meaning of the theory of preventive maintenance unfortunately is not widely understood or well explained.

To the best knowledge of the author, Prof. A.Birolini [3] developed the most comprehensive analysis of preventive maintenance with rigorous check of required assumptions. An objective of this work is to apply this approach in the aviation domain, assuming real-time checking of the aircraft condition and ability of prediction of conditions deterioration.

The preventive maintenance might increase confidence about the aircraft's current state. To achieve this one requires the development of an aircraft model as well as model for estimating an impact of fault on the system. One has to take into account an estimation of efficiency of this implementation.

Challenges in the area of preventive maintenance are:

- Dependence of the periods of preventive maintenance on parameters and data.
- Role of checking and testing coverage on quality parameters.
- Development of generalised model including these two factors.

The last bullet point deals with efficiency of processing of flight data and evaluation of system condition *pre, during,* and *post* mission. Then preventive maintenance development is based upon:

- Introducing of PACC.
- Development of a model for preventive maintenance based on conditional probability.
- Reasoning and inference about assumptions of preventive maintenance.
- Analysis of main factors that influence on the period of preventive maintenance.
- Evaluation of an impact that PACC has on the policies of preventive maintenance.

Some criteria for judging PACC success are:

- How big is a gain of PACC in comparison with classic preventive maintenance?
- Can PACC allow varying periods of maintenance as a function of a condition of an aircraft proven/evaluated/estimated during flight, using flight data processing?
- Can PACC's real-time ingest and analysis, provide finer grained fidelity to in-flight system health inferenceing and to post flight cause-effect analysis?
- What level of mission reliability can realistically be achieved?

It is certain that full coverage of all possible faults of the complex systems *cannot* be achieved in practice. It is also certain that 100 % level of confidence of estimations of aircraft conditions *cannot* be guaranteed. So, how far can we go here? Can we provide clear and substantial coverage of faults and define trends especially the most dangerous ones that lead to accidents? How does a PACC implementation define or change the period of preventive maintenance? Can PACC support required maintenance by location of possible faults, and does it reduce the overall inspection time? It is at least intuitively clear that implementation of PACC increase flight safety and aircraft reliability. However, justification of the gain might be required to achieve economic efficiency of a PACC implementation.

PACC, Conditional and Preventive Maintenance

Preventive maintenance estimations deal with processes of system degradation due to wear and tear, i.e. due to ageing of materials and the effects of utilisation. Purpose of *conditional maintenance* is to detect hidden faults and to anticipate latent faults to avoid their occurrence in a timely way and thus avoid actual fault impact on the system. The so-called *latency* of the fault is a phenomenon of the possible trend of a parameter, which is related to a fault (or faults). *Latency* also might have another reason, caused by erroneous decoding of a fault. This happens when the aircraft or vehicle is used in limited modes of flight and/or recorded parameters and variables are not representative, etc.

Let us consider an aircraft as a repairable structure with periodic maintenance at T_{PM}, $2T_{PM}$,...; at $t = 0$ consider the aircraft as new. Initially we analyse the aircraft reliability assuming that the elapsed time of periodic maintenance is negligible in comparison with the time of aircraft operation—(quite a realistic assumption as ~ 300 flight hours correspond to ~ 0.5 h of maintenance in commercial aviation, further (CA).

Further research might introduce a non-negligible period of maintenance (PM). There are other factors that influence reliability: repair time, incomplete coverage of testing and quality of maintenance. It might be interesting to investigate more advanced features and assumptions derived for PACC implementation for an aircraft implementation such as *sensitivity* to coverage of testing, reduction of maintenance time due to real time (RT) processing of flight data and growth of maintenance quality. Recent papers [12, 13] cover the role of malfunctions in reliability of the system and initiates research in this direction. Other promising research areas in reliability modelling are:

- The impact of the volume of data on quality of evaluation of vehicle condition.
- Time of processing of flight (current) data.
- Reliability vesus models available ("are the structure models available good enough?").
- The impact of flight data on safety ("how much we need to know to be safe?").

In data dependencies further areas of required research are:

- The relationship between accumulated and current flight data to define condition.
- Data integrity in the long term (distillation of flight data trends).
- The efficiency of data access for evaluation of conditions according to PACC.

Organizationally, a better policy of maintenance can be developed if the fundamental model includes in its implementation plan, the introduction of support for unavoidability of maintenance procedures and spreading the cost of maintenance. Both features should be considered for maintenance policies with and without PACC implementation. This research is also might be helpful in convincing insurance companies to revisit current policies existing at the aircraft and similar markets.

Conditional Maintenance

Let us assume that maintenance takes negligible time, relative to the operational life of the aircraft. Four options are possible here:

1. PM is not performed and the aircraft is considered *as good as new*.
2. PM is not performed and the aircraft is considered as non-suitable for further flights (e.g. because some resource necessary for flight is exhausted).
3. As a result of testing procedures the aircraft is considered not to be flight worthy (due to insufficient test completeness or test trustworthiness) and PM is not performed.
4. The aircraft is considered to be potentially not flight worthy and PM is performed instead of a full-scale repair.

The fourth assumption is now explored. Ideal maintenance assumes that at times 0, T_{PM}, $2T_{PM}$,... the system (aircraft) is 'as good as new'. The reliability function for the aircraft without preventive maintenance is:

$$R(t) = 1 - F(t) \text{ for } t > 0, \ R(0) = 1 \tag{15.1}$$

where F(t) is the distribution function of the failure-free operating time of a single item structure and, for simplicity, it is assumed that it is represented by the exponential distribution $F(t) = 1 - e^{-\lambda t}$ in the period t, and λ is constant. Introducing conditional maintenance changes the form of the reliability function for the aircraft as follows:

$$R_{PM}(t) = R^n(T_{PM})R(t - nT_{PM}) \quad \text{for } nT_{PM} < t \leq (n+1)T_{PM} \text{ and } n = 0, 1, 2, \ldots \tag{15.2}$$

R(t) and R_{PM}(t) give the probability for no failures (faults) in the period (0, t), without and with ideal maintenance.

If an aircraft is considered as a system without maintenance and repair then its reliability in its simplest form (assuming a constant failure rate λ) can be presented by the reliability function given by (15.3):

$$R(t) = e^{-\lambda t} \tag{15.3}$$

$R(t)$ per Eq. (15.3) is depicted in Fig. 15.3, with $\lambda = 0.3$ and time parameter $t = [0 \ldots 10]$. Figure 15.2 solid line is R(t), dashed line is threshold R_o. Threshold 0.2 was chosen very low to increase visibility. The dot-and-dash line marks the point where R_o is reached the system condition when aircraft or system should be put out of service.

The threshold R_o (straight line) represents the minimum level of system reliability required to continue safe operation. For this example, $R_o = 0.2$ (chosen particularly low to increase visibility), the reliability approaches the threshold R_o at time 5.4. Aircraft in modern management schemes should be serviced when aircraft condition reaches a certain level. This approach is known as *conditional*

Fig. 15.3 Reliability function R(t) for the case of constant failure rate λ

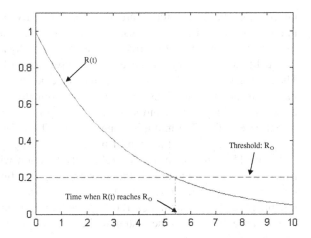

maintenance. Usually evaluation of conditions of aircraft after maintenance is overoptimistic and assumes, in particular, that maintenance fixes all possible faults in the aircraft. This makes it possible to set maintenance procedures periodically, at times when the model shows that reliability is reaching the point when maintenance is necessary and considering an aircraft as good as new after maintenance.

Note that assumptions of ideal conditional maintenance and threshold level of reliability allowed are combined to define the size of intervals between maintenance activities. Existing practice tends to set maintenance intervals to be equal. Formally, the *reliability function* $R_{PM}(t)$ with ideal conditional maintenance is based on the following assumptions:

Assumption 1 100 % coverage i.e. maintenance restores the system completely

Assumption 2 The interval between two successive maintenances is constant: T_{PM}

Assumption 3 Maintenance is produced instantly and does not delay the usage schedule

In such a situation, it is possible to consider a *mission reliability MR(t)* as reliability function between two successive periodic maintenance actions, i.e. with t starting by 0 at each maintenance phase. For the case of constant failure rate λ this leads to (see Fig. 15.3).

$$MR(t) = e^{-\lambda(t-nT_{PM})}, \quad \text{for } nT_{PM} < t \le (n+1)T_{PM}, \ n = 0, 1, 2, \ldots \quad (15.4)$$

It is also possible to consider $MR_n(t)$ and assign the mission reliability to the corresponding mission. As stated above, it is assumed that periodic ideal conditional maintenance restores the system to the state 'as good as new'. The approach is well known in aviation and other safety critical industries as it enables reliability theory to be applied for estimation of conditions of the system over life cycle of operation. Note here that this kind of reliability models is quite optimistic and can, at best, be used as a guide: firstly intervals between maintenance inspections are rarely equal because aircraft are now used heavily e.g. in *chain flights*, with interval between flights less than 1.5 h; secondly, commercial aviation suffers from sporadic and far from perfect maintenance; thirdly as shown in [1] and above, the quality of regular maintenance across all segments of aviation is far from ideal. The main causes for this are a) the maintenance personnel, and b) lack of objective models to define conditions of aircraft. Additionally, latent aircraft faults often exist quite a long time: from some minutes up to several years see for example recent case with A380 multiple wing defects). Therefore, more realistic assumptions are required for estimation of mission reliability.

Figure 15.4 presents a mission reliability function with ideal periodic maintenance, where the solid curve is the mission reliability function, the dashed bottom line is the acceptability threshold, and the dot-and-dash line indicates the perfectly reliable state of the system, i.e., 100 % reliable. It is assumed full coverage of ideal maintenance that returns the system to the state 'as good as new', and maintenance periods are: T_{PM}, $2T_{PM}$,...,nT_{PM}.

Conditional Maintenance with Incomplete Coverage

Regretfully, the optimism of existing declarations about the quality of maintenance and complete coverage of the system faults has short lived: in November 2010 alone aircraft accidents with A380 and Boeing 747 and A380 2012 multiple wings mishaps show that *coverage* is far from required level. Denote *coverage* as α, $\alpha < 1$. The *mission reliability function* assumptions are formally presented below for the case of *maintenance with incomplete coverage*:

Assumption 1 Coverage is not 100 %. Coverage percentage is 100 α%, where $0 < \alpha < 1$, and is assumed to be constant over all maintenance actions

Assumption 2 Maintenance is instantaneous and doesn't delay aircraft schedule

Assumption 3 A threshold MR_0 of acceptable mission reliability is given (fixed)

Assumption 4 T_{PM} is a function of several variables, including α, λ and MR_0

Fig. 15.4 Mission reliability with ideal preventive maintenance

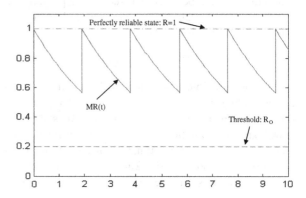

Mission reliability is then calculated according to:

$$MR(t) = \alpha^K e^{-\lambda\left(t - \sum_{i=0}^{n} T_{PM(i)}\right)}$$

$$\text{for } \sum_{i=0}^{n} T_{PM(i)} < t \le \sum_{i=0}^{n+1} T_{PM(i)}, T_{PM(0)} = 0 \text{ n} = 0, 1, 2\ldots \tag{15.5}$$

The resulting mission reliability curve for this case is presented in Fig. 15.5. Equation (15.5) is in particular true for $\alpha \approx 1$. Note that system is as good as new after the n-th PM and that as well a n restart by 0 at each corrective maintenance yielding system as good as new. It is now assumed that maintenance takes place when the system (an aircraft) reaches the threshold reliability i.e. when:

$$MR(t) = MR_0 \tag{15.5a}$$

This case has some theoretical interest, as it might be useful to analyse the role of all the variables that define behaviour of period of maintenance T_{PM}.

Calculating $T_{PM}(i)$, for $i = 1,2,\ldots,n$, and taking into account the role of the other variables such as MR_0, α and λ; then $T_{PM}(i)$ is given as:

$$T_{PM}(i) = \frac{1}{\lambda} \ln \frac{\alpha^{i-1}}{MR_0}, \quad i = 1, 2, \ldots \tag{15.6}$$

This model is more realistic, enabling to schedule maintenance when the system (aircraft) reaches the threshold of acceptable mission reliability. Observe here that the interval between successive maintenance inspections $T_{PM}(i)$ is shrinking significantly along life cycle of aircraft operation. The relative decrease can be evaluated by the rate of decrease of $\Delta T_{PM}(i)$:

Fig. 15.5 Conditional periodic maintenance with incomplete coverage

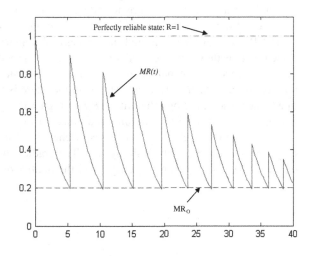

$$\Delta T_{PM(i)} = \frac{T_{PM(i)} - T_{PM(i+1)}}{T_{PM(i+1)}} \tag{15.6a}$$

or, by the function of the interval index:

$$\Delta T_{PM} = \frac{T_{PM(i)} - T_{PM(i+1)}}{i} \tag{15.6b}$$

Figure 15.5 presents the function of mission reliability for the case of periodic maintenance with incomplete coverage. The solid curve is the mission reliability curve, the dashed line is the threshold, and the dot-and-dash line indicates the perfect reliable state of system, i.e. as if 100 % reliable. It is assumed that while the threshold is reached, maintenance is carried out. But for this example, because of incomplete coverage, the mission reliability of the system cannot return to 100 % after maintenance, and the amplitude of recovery of conditions after iterations of maintenance gradually degrades over time.

The actual condition of aircraft varies between thresholds MR_o and $MR(t)$ between two successive maintenances. When mission reliability approaches MR_o it should be grounded in the interests of safety. Maintenance period shown with picks defined by T_{PM}, $2T_{PM}$, $3T_{PM}$,... etc.

Maintenance with Implementation of PACC

PACC introduces a new CPS process in aircraft management: on-line checking of the aircraft's condition. On-line checking is a process of real-time (during the flight) checking of the aircraft's main elements, including hardware (in general), electronics and pilot. The aim of checking is detection of degradation or change in behaviour and, when possible, recovery of the suspected element or subsystem, conserving the system's reliability and safety. When recovery is not possible the preventive nature of PACC aims to reduce the level of danger, risk etc.—aiming for graceful degradation of an object or service quality to the object's users.

The Process of Checking and the Process of Maintenance are independent in principle; thus they can be considered as concurrent processes as well as sequential ones. The checking or maintenance activities can be started when required, when possible or just when convenient. The main idea here is to carry out checking well in advance when mission reliability $MR(t)$ is higher than threshold reliability MR_0, making degradation of the aircraft conditions during flight less probable.

When applied together the processes of checking and conditional maintenance may increase the reliability of the system. The gradient of this change is a function of the quality of checking (coverage) and the quality of maintenance.

For consistency of analysis of the impact of PACC implementation we introduce following conditions:

- A constant failure rate.
- Maintenance is not ideal and coverage is less than 100 %.
- Minimum acceptable reliability threshold is introduced as before.

Some other assumptions relate to the checking process. Formally, the mission *reliability function* for preventive maintenance with an introduced online checking process is based on the following assumptions:

Assumption 1 Coverage of maintenance is not ideal. Coverage of maintenance is $\alpha_M 100$ %, where $0 < \alpha_M < 1$, and α_M is assumed as a constant

Assumption 2 Threshold MR_0 exists for $MR(t)$

Assumption 3 Online checking process is introduced. The period for checking is T_{PC} and T_{PC} is a constant

Assumption 4 The system can dynamically scale. Thousands of checks may have to occur within different time intervals. The resource processing pool is tuned via scalable processes to keep T_{PC} a constant per each required check

Assumption 5 After each online checking, the confidence about the system's conditions is increased, therefore $MR(t)$ is also increased, and this confidence is $\alpha_C 100$ %, while $0 < \alpha_C < 1$ and α_C is a constant

Assumption 6 The period between two successive maintenance inspections is $T_{PM}(i)$. $T_{PM}(i)$ is a variable, actually a function of i, R_0, α_C, α_M, λ and T_{PC}

The mission reliability function (rigorously speaking conditional probability of absence a failure in the previous checking period as it is clarified below) for an aircraft is then calculated according to:

$$MR(t) = MR_i \alpha_c^n e^{-\lambda(t-nT_{PC})}, \quad \text{for n } T_{PC} < t \leq (n+1)T_{PC}, \text{ n} = 0,1,2,\ldots \quad (15.7)$$

For MR(t) in Eq. (15.7) n stands for the n-th on-line checking period. For a new system, $MR_0 = 1$. MR_i follows from Eq. (15.5) as

$$MR_i = \alpha_M^i, \quad i = 0,1,2 \qquad (15.7a)$$

where *i* corresponds to the *i*th maintenance period, MR_i denotes the initial value of *mission reliability* at the beginning of *a maintenance period*, $MR_i \alpha_c^n$ denotes the initial value at the beginning of *an online-checking period* respectively. Note that n in Eq. (15.7) start at 0 at each maintenance period;

When the mission reliability of an aircraft reaches the threshold MR_o it should be grounded awaiting for preventive maintenance, so:

$$MR_i \leq MR_0 \qquad (15.7b)$$

From a practical point of view, the online checking period should be constant, as per Assumption 3 above, and the checking procedure should start at the beginning of the following period. Suppose initially that checking takes no time, and maintenance will be carried out instantly. Even if time delay due to the checking process has to be considered, we still assume that the maintenance is carried out only at the end of the following online-checking period. Let index n be the serial number of an online-checking period, and index i be the serial number of a maintenance period. The online-checking period T_{PC} and the maintenance period $T_{PM}(i)$ relates as:

- The online-checking period T_{PC} is a constant, the maintenance period $T_{PM}(i)$ is a variable.
- $T_{PM}(i)$ contains a certain number of T_{PC}.

With these assumptions mission reliability per Eq. (15.7) is shown on Fig. 15.5.

Figure 15.6 is an example of a mission reliability function under conditional maintenance with on-line checking, where the solid curve is the mission reliability curve, the dashed line is the threshold, and the dot-and-dash line indicates the perfect reliable state of system, i.e., 100 % reliable. As shown on Fig. 15.6, once an on-line checking period arrives, the latest system states are measured and analysed.

After each online-checking process the latest system states are available and, therefore, the awareness and confidence about the system both recover a little bit (subject to no faults being detected), so does the mission reliability curve. When the mission reliability reaches the threshold, maintenance is carried out just as with preventive maintenance in Fig. 15.5. The rate of mission reliability degradation is a topic for further investigation, searching for the ways to slow down a system degradation using ICT technologies.

When no maintenance is scheduled for a long time (the actual situation in commercial and general aviation) the mission reliability of an aircraft will reach

Fig. 15.6 Preventive maintenance with on-line checking

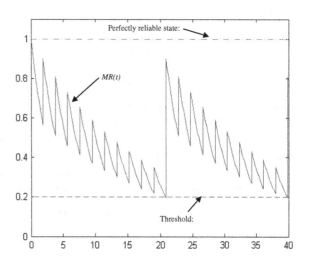

the threshold MR_o. The rate of mission reliability with on-line checking in fact decreases slightly faster, due to added unreliability of checking system itself. Checking with subsequent maintenance, on the contrary, increases mission reliability. The gap of confidence between a point in time *before* checking and *after* the checking will from now on be referred as a *corridor of mission reliability*.

The Mission Reliability Corridor: Introduction and Definitions

The basic model of a mission *reliability corridor* δ is defined using practical assumptions and a set of scenarios as in the previous sections.

Suppose no serious system faults occur, and then the mission reliability corridor is defined as the safe operational area where the curve is normally expected to stay under the online-checking scheme. The corridor defines the value that mission reliability curve could reach in each on-line checking period, and, therefore, corridor effectively helps to decide when to carry out maintenance in order to avoid violating the given threshold. On the other hand, the 'width' of the mission reliability corridor will help to define the requirements for software and hardware of the system that perform conditional control. Prediction or estimating of system condition depends on volume of data, complexity of a model used and performance of hardware, all integrated into allowable or not time delays. The corridor is plotted in Figs. 15.7, 15.8, 15.9, 15.10, and 15.11 and represented as dotted lines.

Definition 1 In each online checking period, the width of the corridor δ is a constant and time independent. During the n-th online checking process a mission reliability corridor $\delta(n)$ is a function of n with width and given as:

$$\delta(n) = MR(nT_{PC}) - MR((n+1)T_{PC}) \tag{15.8}$$

Clearly the corridor under this definition becomes too conservative at the end of each online checking period; the cause is that the amplitude of coverage by on-line checking shrinks as time goes on, as illustrated in Fig. 15.7.

In other words, the upper boundary $\delta U(n)$ and the lower boundary $\delta L(n)$ of the mission reliability corridor in Fig. 15.7 are given respectively given as:

$$\delta_U(n) = MR(nT_{PC}) \tag{15.8a}$$

$$\delta_L(n) = MR((n+1)T_{PC}) \tag{15.8b}$$

In Figs. 15.7, 15.8, 15.9, 15.10, and 15.11, the solid plot line is the mission reliability curve, the dashed line is the threshold level, and the dot-and-dash line is the initial reliability level. The dotted lines around mission reliability curve show the corridor, and the vertical dotted lines indicate online-checking periods.

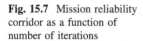

Fig. 15.7 Mission reliability corridor as a function of number of iterations

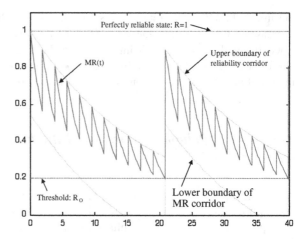

Definition 2 A time-varying corridor with the width δ varies over time within each online checking period. For the n-th online checking process $\delta(t)$ is given as:

$$\delta(t) = MR(nT_{PC})\alpha_C^{(t-nT_{PC})/T_{PC}}\left(1 - e^{-\lambda T_{PC}}\right), \quad nT_{PC} \leq t < (n+1)T_{PC} \qquad (15.9)$$

Actually, $MR(nT_{PC})\alpha_C^{(t-nT_{PC})/T_{PC}}$ in Eq. (15.9) defines the upper limit of the corridor at time t. Assume a hypothetic system with mission reliability of the same value at the upper limit of the corridor at time t, then $MR(nT_{PC})\alpha_C^{(t-nT_{PC})/T_{PC}}\left(1 - e^{-\lambda T_{PC}}\right)$ is the mission reliability after an online checking period T_{PC}. The width of the corridor δ at time t, $\delta(t)$ equals the difference between the upper limit of the corridor at time t and the reliability of a system at time $t + T_{PC}$. It is evident that the width of corridor varies over time.

Fig. 15.8 Mission reliability corridor as a function of time

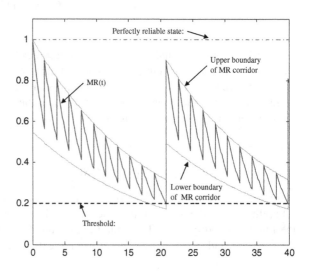

Fig. 15.9 On-line checking performance requirement—β gap

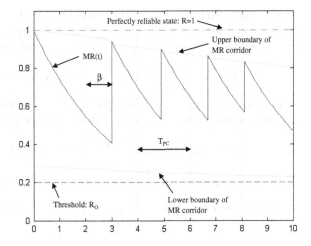

Fig. 15.10 Mission reliability with calculation after the checking period

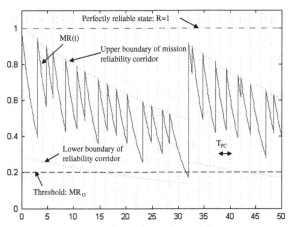

Fig. 15.11 Mission reliability with checking for reaching boundary

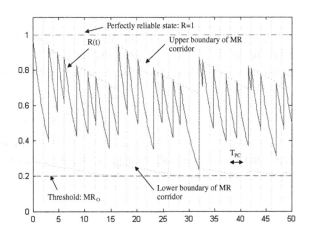

The corresponding corridor of the reliability curve is illustrated in Fig. 15.8. Note that it shrinks with the amplitude of coverage by on-line checking.The width of the reliability corridor in Fig. 15.8 is given as follows:

$$\delta(t) = R(nT_{PC})\alpha_C^{(t-nT_{PC})/T_{PC}}\left(1 - e^{-\lambda T_{PC}}\right), \quad nT_{PC} \leq t < (n+1)T_{PC}. \quad (15.9a)$$

In other words, the upper boundary $\delta_U(n)$ and the lower boundary $\delta_L(n)$ of the mission reliability corridor in Fig. 15.8 are given respectively as:

$$\delta_U(t) = R(nT_{PC})\alpha_C^{(t-nT_{PC})/T_{PC}}, \quad nT_{PC} \leq t < (n+1)T_{PC} \quad (15.9b)$$

$$\delta_L(t) = R(nT_{PC})\alpha_C^{(t-nT_{PC})/T_{PC}} e^{-\lambda T_{PC}}, \quad nT_{PC} \leq t < (n+1)T_{PC} \quad (15.9c)$$

Clearly, this corridor is much less conservative than introduced by Definition 1.

Defining the Frequency of the On-line Checking Process

Assumption 1 Online checking process starts at the beginning of each period of use.

Figure 15.9 illustrates impact of time required for real time data processing on mission reliability, where the dotted lines are used to indicate each on-line checking period, which in this case is set as 2-time-units long. Because the measurement and analysis of the latest system states can not be completed immediately at the beginning of each on-line checking period, the awareness and confidence about the system are not improved until these data are available, and therefore there is a delay β on the coverage of the mission reliability curve in each online checking period. So β is the time required for data processing, which may vary, and has an upper bound β_{max}, i.e., $\beta \leq \beta_{max}$. The worst case should be:

$$\beta_{max} = T_{PC} \quad (15.10)$$

The question is, what is the influence of a data processing delay on the definition of the corridor, i.e. the impact of β_{max} on $\delta(t)$, assuming the second definition of a corridor is adopted? When β_{max} is taken into account, $\delta(t)$ should be calculated by:

$$\delta(t) = MR(nT_{PC})\alpha_C^{(t-nT_{PC})/2T_{PC}}\left(1 - e^{-2\lambda T_{PC}}\right), \quad nT_{PC} \leq t < (n+1)T_{PC} \quad (15.11)$$

Compared with "T_{PC}" in Eq. (15.9), "$2T_{PC}$" in Eq. (15.11) embodies the maximum delay due to online data processing, in the case that β_{max} is almost out of synchronization with T_{PC} in its period.

Avoiding R_0 Being Violated in the Corridor When Delay Occurs

Implementation of principle of active conditional control requires that mission reliability should not fall below the threshold R_0 even in when β_{max} is taken into account. This could be achieved in one of three methods:

Method 1. Within each online checking process, when data processing is finished, check whether the mission reliability is below the threshold R_0. In this case, due to the delay caused by data processing, the threshold could be violated. Figure 15.10 shows that when online checking is carried out at time 30 the mission reliability is above the threshold but then goes below the threshold when the online checking process is finished at time 32.

Method 2. In each online checking process, check whether the bottom line of the corridor is below the threshold R_0, i.e.:

$$MR_I \alpha_C^{(n-n_{AM})} \alpha_C^{rem(t,T_{PC})} - \delta(t) \leq R_0 \qquad (15.11b)$$

where the first term of the relation defines the top of the corridor, and "rem" signifies the remainder after dividing t by T_{PC}. The result of applying this method is illustrated in Fig. 15.11. The maximum delay, i.e. T_{PC}, is taken into account when defining the width of corridor in (Eq. 15.11) so that the mission reliability is always within a corridor even when there is data processing delay. Consequently, the mission reliability in Fig. 15.11 never reaches the lower threshold because maintenance is carried out in time before the bottom of corridor touches the threshold.

Method 3. Define a buffer zone, i.e. $[MR_0, R_B]$ then in each online checking process, check whether the mission reliability is within the buffer zone, i.e.,

$$MR((n+1)T_{PC}) \leq MR_0 + MR_B \qquad (15.11c)$$

The result of introducing a buffer zone is illustrated in Fig. 15.12, where the buffer zone is represented as the area between the dashed line and the dot-and-dash line. Due to the delay caused by online data processing there is a possibility that the reliability will 'enter' the buffer zone. Once this happens, maintenance must be carried out in order to avoid the reliability going further below the threshold.

Maintenance Versus PACC

Previous sections show that preventive maintenance with PACC is more efficient than known conditional or preventive maintenance approaches. The quantitative analysis might help to see how much. Comparisons might be performed using time between two successive maintenance sessions, the lifespan of the system under a certain maintenance strategy, and how many times maintenance is carried out

Fig. 15.12 Mission
reliability with checking
within a buffer zone

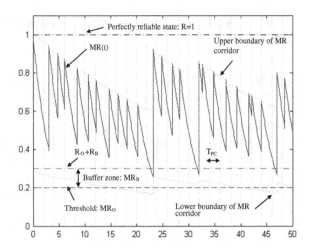

during the life time of system. But here we propose an integration of mission
reliability over a given time period, i.e. the volume of the area encircled by the
mission reliability curve and the reference axes. A main reason for this index is to
compare schemes of conditional control and preventive maintenance as introduced
above.

The integration values of mission reliability under conditional maintenance and
preventive maintenance with PACC are calculated by Eqs. (15.12), (15.13)
respectively:

$$V_{CM}(T_1) = \int_0^{T_1} MR_{CM}(t)dt, \tag{15.12}$$

$$V_{PM}(T_2) = \int_0^{T_2} MR_{PM}(t)dt, \tag{15.13}$$

where $MR_{CM}(T)$ and $MR_{PM}(T)$ are given by Eqs. (15.3) and (15.5).

Then efficiency of the preventive maintenance with PACC over conditional
maintenance can be assessed as:

$$y(T_1, T_2) = \frac{V_{PM}(T_2) - V_{CM}(T_1)}{V_{CM}(T_1)} \tag{15.14}$$

Let us assume $T_1 = T_2$. This means we compare the mission reliability of
system with preventive maintenance with PACC with the one with conditional
maintenance in a same period of time. Figure 15.13 gives an example of such a
comparison, where $T_1 = T_2 = 40$.

For Eqs. (15.12) and (15.13): $V_{CM}(40) = 15.5961$, $V_{PM}(40) = 18.5084$ and
$Y(40) = 0.1867$.

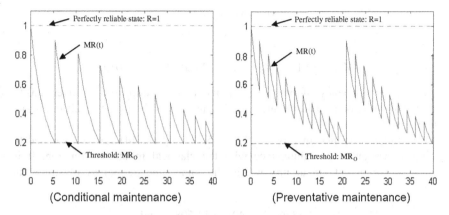

Fig. 15.13 Efficiency of conditional and preventive maintenance with PACC

$V_{PM}(40) > V_{CM}(40)$ means that in the specified 40 unit time period the system under preventive maintenance with PACC has a higher reliability, in other words, the efficiency of preventive maintenance using PACC is about 20 % better compared with conditional maintenance. Accordingly Fig. 15.13 preventive maintenance with PACC could increase period between two sequential maintenance sessions, therefore overall cost of maintenance for a vehicle reduces.

Let T_1 and T_2 be the lifespan of the system under preventive maintenance with PACC and conditional maintenance, respectively. Then the value of y in Eq. (15.14) can be used to assess how much extra reliability the adoption of preventive maintenance has created relative to a conditional maintenance scheme.

Comparison of the left and right boxes of Fig. 15.14 shows that the conditional maintenance system will no longer be able to recover after point 44.6 in time, while under the preventive maintenance with PACC, the critical time is 129.1. One can then easily deduce from Eqs. (15.12) and (15.13) that:

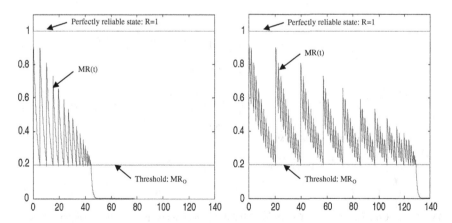

Fig. 15.14 Comparison of efficiency of conditional and preventative maintenance with PACC

$$V_{CM}(44.6) = 16.6707, \; V_{PM}(129.1) = 50.2670$$

and

$$(V_{PM}(129.1) - V_{CM}(44.6))/V_{CM}(44.6) = 2.0153$$

Thus, the efficiency of preventive maintenance is improved by over 200 % compared with conditional maintenance. Figure 15.14 shows the result in a more intuitive way.

The indexes defined in Eqs. (15.12), (15.13) and (15.14) can be extended to compare preventative maintenance with the classical reliability function. It is worth to compare them at first within the same time period, as illustrated in Fig. 15.14:

$$V_{CRF}(40) = 3.3336, V_{PM}(40) = 18.5084, \; \text{and}$$
$$(V_{PM}(40) - V_{CRF}(40))/V_{CRF}(40) = 4.5521$$

Let us estimate gain in mission reliability for the systems with implemented active conditional control against the standard system for the whole period of functioning. The classical mission reliability function reaches the threshold at the time 5.4 (Figs. 15.15 and 15.16). When preventive maintenance with PACC is applied the mission reliability declines to lower bound much slower—after the time 129.1, and then one has:

$$V_{CRF}(5.4) = 2.6739,$$
$$V_{PM}(129.1) = 50.2670 \; \text{and} \; (V_{PM}(129.1) - V_{CRF}(5.4))/V_{CRF}(129.1) = 17.7991$$

Figure 15.16 illustrates the significant advantage in mission reliability when preventive maintenance with PACC is applied in comparison with the system described by classic mission reliability.

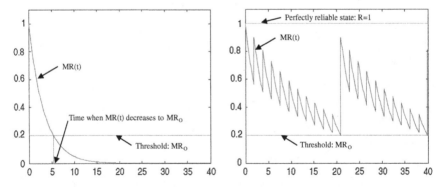

Fig. 15.15 Comparison of the CLASSIC reliability function and preventative maintenance with PACC

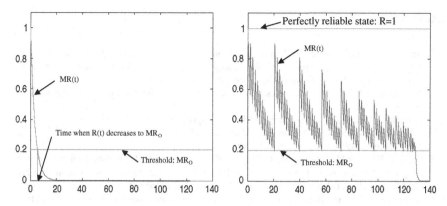

Fig. 15.16   Classical reliability function versus preventative maintenance

Conclusions

- The Principle of Active Conditional Control has been analysed in terms of the mission reliability gain for aircraft maintenance. The Classical, Conditional and Preventative approaches to maintenance have been compared quantitatively.
- Principle of Active Conditional Control assumes continuous application of knowledge of aircraft structure and results of flight data aiming to improve safety and mission reliability of aircraft, the quality of maintenance and reducing the cost.
- Implementation of this principle enables the monitoring of reliability in real time of aircraft application and offers 20–25 % growth of mission reliability.
- Mapping between flight information and aircraft safety or mission reliability, the role and structure of information as well model of aircraft and impact of flight conditions are subject of a special integrated research.
- To benefit from proposed approach an aircraft (as well as any other safety critical real-time system) should be designed introducing principle of active conditional control from the conceptual draft of a system, benefitting from knowledge about dependencies between aircraft elements and subsystems.
- Aviation is the most complex area for the application of technological advances: complex and long working periods, an extremely wide range of operation conditions, multi-disciplinary skills needed from personnel involved. Therefore the Principle of Active Conditional Control and its implementation must become the subject of future multidimensional research to improve aviation safety and efficiency.

Acknowledgements   Author thanks a reviewer of the paper for comments and detailed arguments this helps to improve paper.

References

1. Annex 1 - "Description of Work", Version 3.1, dd. 12/11/2004, for ONBASS. PF6: *"Integrating and strengthening the Europe-an Research Area,* Thematic Priority: Aeronautics and Space, Contract No.: AST4-CT-2004-516045
2. Appendix   F   http://science.ksc.nasa.gov/shuttle/missions/51-l/docs/rogers-commission/Appendix-F.txt
3. A.Birolini, Reliability Engineering, 6th Edition, Springer Verlag, 2010.
4. CAD 418 Condition Monitored Maintenance: an Explanatory Handbook http://www.cad.gov.hk/english/pdf/CAD418.pdf
5. Galie T, Roemer M, Gregory M, J. Kacprzynski J, Byington C, Prognostic Enhancements to Diagnostic Systems for Improved Condition-Based Maintenance 1 http://www.dtic.mil/cgi-bin/GetTRDoc?AD=ADA408880&Location=U2&doc=GetTRDoc.pdf
6. Huang Gary K, Lin Kuen Y, A Method for Reliability Assessment of Aircraft Structures Subject to Accidental Damage, depts.washington.edu/amtas/publications/presentations/Lin_AIAA_4-05.pdf
7. Kingsley-Jones Max, Reliability lessons learned. In-Service Report:A340-500/600 Flight International, 3-9 May 2005 pp34-39
8. Middleton D.H. Aircraft Maintenance Management Part 3, Aircraft Engineering and Aerospace Technology, Year: 1993 Volume: 65 Issue: 2 Page: 6 – 9, ISSN: 0002-2667, DOI:10.1108/eb037340, Publisher: MCB UP Ltd
9. Part VI - General Operating and Flight Rules *Canadian Aviation Regulations 2009-2* Standard 625 Appendix B - Maintenance Schedules http://www.tc.gc.ca/civilaviation/regserv/affairs/cars/part6/standards/a625b.htm
10. Summerfield J.R A Model for Evaluating Fleet of Transport Aircraft, Logistic Department The RAND Corp 12 Jan 1960 http://www.rand.org/pubs/papers/2009/P1882.pdf
11. Schagaev I., Concept of Dynamic Safety for Aviation ISSC 1998, Seattle, USA.
12. Schagaev I., Redundancy Classification and Its application for FT Computer System Design IEEE TESADI-01, Arizona, Tucson, October 2001.
13. Schagaev I., Reliability of Malfunction Tolerance, Proc. of the International Multi-conference on Computer Science and Information Technology, pp. 733–737, ISBN 978-83-60810-14-9
14. http://www.it-acs.co.uk/files/GB2448351B.pdf
15. http://www.it-acs.co.uk/files/new_challenges.pdf
16. Suh, N., Complexity Theory and Applications, Oxford University Press, 2005
17. Lee, T., Axiomatic Design & Complexity Theories: Information Axiom, Lecture 2006

Chapter 16
Long Range Wireless Data Acquisition Sensor System for Health Care Applications

Abdullah Eroglu

Introduction

In today's health care environment, it is common to have patient connected to several medical devices simultaneously. These devices can be responsible for monitoring several critical patient parameters such as body temperature, heart rate, and blood pressure. Medical professionals who are in close proximity to these devices get the information provided by various devices and take the necessary action needed such as adjusting the dose of the drug delivered or regulate the breathing. Patient's life supporting units become totally dependent on medical professionals. Humans are subject to fatigue, miscommunication, distractions, and misinterpretation of the information, and several other factors. These can produce an undesirable patient outcome. As a result, the real-time monitoring of the information related to critical patient parameters carries great importance for patient health. The need to provide health care services in this era is to measure and collect patient's critical parameters using wireless technology and analyze the information and control vital medical devices connected to patient remotely in real-time. This allows continuous monitoring of the patient and instant adjustment of devices when needed. It allows the delay and removes the possible errors due to human factor. Remote measurement and collection of the patient data, its communications and analysis depends several factors. The measurement of the critical patient parameters can be done using wearable or implementable sensors. Over the past several decades, sensors have become essential components in communication systems and they become very important recently in medical applications [1–4]. Sensors transform the physical quantities such as temperature, flow rate, acceleration, etc. into electrical signals that serve as inputs for control systems [5]. Sensors can be active or passive depending upon whether or not there is an

A. Eroglu (✉)
Indiana University–Purdue University Fort Wayne, 2101 E. Coliseum Blvd,
Fort Wayne, IN 46805, USA
e-mail: eroglua@ipfw.edu

S. C. Suh et al. (eds.), *Applied Cyber-Physical Systems*,
DOI: 10.1007/978-1-4614-7336-7_16,
© Springer Science+Business Media New York 2014

on-board battery supply. Active sensor platforms are available in different forms such as Berkeley Motes [6] or BTnodes [7]. Passive sensors need an external power source to power their on-board electronics [8–10]. The communication of the data collected from the patient with sensors can be transmitted using RFID technology. RFID is a method to store and retrieve data using RFID tags/ transponders remotely [11]. It an effective technique in track and trace applications [12, 13]. There are challenges associated with RFID technology including communication distance, cost [14], RFID communication protocols [15] and security measures [16]. These wireless systems allow non-contact reading and are excellent for use in medical applications. Conventional RFID system includes tags, reader, local software/infrastructure and back end system. RFID tag is the identification device in the system and contains at least a microchip attached to an antenna that sends data to an RFID reader. RFID tags can be active, passive or semi-passive tags. There are two important issues that stand as challenge in use of sensors and RFID systems in medical applications. One of the important issues is the interface between sensors and RFID systems. The second critical issue is the limited reading distance that existing RFID systems have. The reading distances are up to 3 ft and 20 ft for high frequency (HF) and ultra high frequency (UHF) RFID systems, respectively.

In this chapter, long range and low cost wireless data acquisition (WDAQ) sensor system for health-care applications has been designed, simulated, and implemented. Single receiver and transmitter pair is used instead of conventional multiple input multiple output antenna systems with sequential logic that reduces the cost significantly. Physical system that has been implemented is composed of three stages: sensor measurement, local communication station and base station. In the first stage, several critical patient parameters are measured using sensor enabled RFID system. The information collected from patient is transmitted to the second stage where local communication station is located. Local communication station gets RFID signal using reader and converts that to dc signal via envelope detector. Output of envelope detector is digitized via A/D converter and sent to stamp microcontroller for processing. The output of microcontroller is interfaced with RF transmitter via RS-232/RS-485 communication. RF transmitter transmits this signal to the third stage where base station is located within 20 mile radius. Base station has RF receiver, which is interfaced with the second microcontroller for post signal processing. Output of microcontroller is fed to D/A converter, which is then sent to SC-2345 signal conditioning unit. The conditioned signal is then acquired by data acquisition hardware DAQPad-6020E and data analysis is performed by LabView software and results are displayed to health care professional. This system gives real-time remote monitoring ability for patients and elders and as a result plays vital role for home care facilities and health care institutes.

Hardware and Implementation of WDAQ Sensor System

The main block diagram of the long range WDAQ sensor system for health-care applications is shown in Fig. 16.1. Stage 1 and Stage 2 in the health-care room can be implemented as illustrated in Fig. 16.2. Wearable sensors and sensor enabled RFID tag are positioned on the human body for measurement of the physical variables. MLX90129 is chosen to be a sensor enabled RFID device in the system. It has an ability to log the sensor outputs when it is operated in the battery mode and transmit them via RFID antenna. It can be interfaced up to four sensors including its internal temperature sensor. This constitutes the Stage 1 shown in Fig. 16.1. MLX90121 is the Reader that can be used with MLX90129. It needs a matching network (MN) for better performance and external supply as illustrated in the figure. Reader is located within the range of the RFID antenna in the room and interfaced with the circuitry to do the following tasks:

- Capture the RFID signal
- Convert it to dc signal
- Amplify the signal for processing
- Convert the amplified signal to digital signal
- Send the digital signal to stamp microcontroller
- Interface output of the microcontroller with RF Transmitter via serial driver
- Transmit RF signal to base station

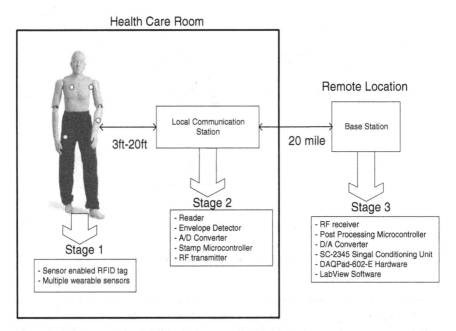

Fig. 16.1 Main block diagram of the long-range WDAQ sensor system for heath care applications

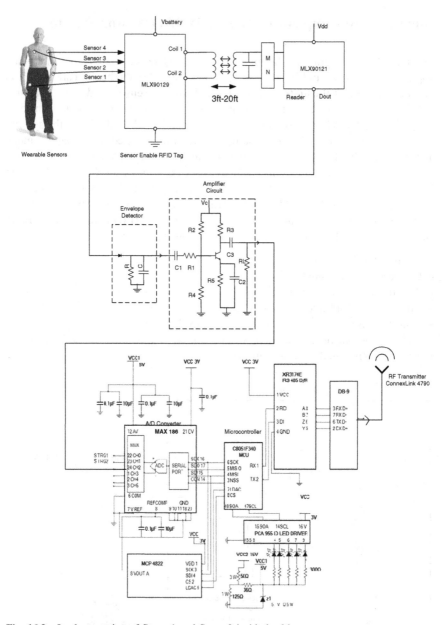

Fig. 16.2 Implementation of Stage 1 and Stage 2 inside health-care room

Envelope detector and amplifier circuit are the circuits designed to feed the signal to MAX 186 which is a 12-bit, 8 Channel single-ended A/D converter. A/D converter's output is interfaced with low cost stamp microcontroller, C8051F340 via serial communication. Microcontroller's output is connected to RF transmitter, ConnexLink 4790 [17] via XR3174E RS-485 driver. The schematics showing all the circuitry including Reader and RF transmitter in the health-care room constitutes Stage 2 as shown in Fig. 16.1. The base station position where Stage 3 and its components are placed are 20 mile away from the location of Stage 2.

The signal that is transmitted by RF transmitter is captured by ConnexLink 4790 RF receiver at the base station. According to the base band signal modulation, digital signal transmission is preferred because digital signal transmission technology has high anti-jamming ability and reliability. The received signal is sent to receiver microcontroller, C8051F340 via XR3174E RS-485 driver for properly extracting the signals and then converted to analog signal using 12-bit, dual voltage D/A converter, MCP 4822. The converted signal is then fed signal conditioning unit where signal conditioning tasks such as filtering, amplification, etc., of the signal take place before it is sent to data acquisition (DAQ) hardware, DaqPad 6020E. LabView software is used to communicate with DAQ hardware and process the received measurement signals to determine the physical variable that is being measured on human body using sensors. The implementation of Stage 3 is shown in the schematics illustrated in Fig. 16.3. It has to be noted ConnexLink 4790 is used for two-way communication since it is RF transceiver system in essence. This gives an advantage to use receiver as transmitter and transmitter as receiver for sending acknowledgment signal and control of medical devices that are connected to patient. Its frequency band changes between 902 to 928 MHz. It features Frequency Shift Keying (FSK) modulation and demodulation capability and it has transmission bit rates up to 115.2 Kbps. Sensitivity of the module at full RF Data Rate is −99 dB. In addition, use of XR3174E RS-485 driver for serial communication is optional. RS-232 line driver can also be used conveniently for serial communication.

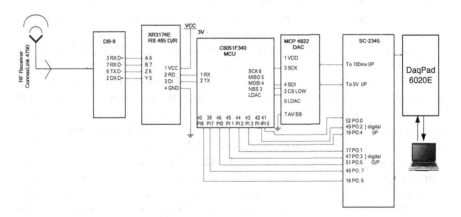

Fig. 16.3 Implementation of Stage 3

Description of the System Operation

The operational principle of the communication system that is designed and implemented in Figs. 16.2 and 16.3 can be illustrated as a block diagram as shown in Fig. 16.4. In this implementation, multiple data signals are sent through only one RF channel using sequential logic. The transmission circuit is controlled by microcontroller C8051F340 that selects the signal to be processed by digital to analog converter (DAC).

The microcontroller (MCU) either cycles through each input channel, or can cycle through only the input channel selected externally. Once the conversion is completed, the digitized value is transferred to the microcontroller's memory. The address of the multiplexer (MUX) selecting the signal is appended to the signal's value. This combined data packet is then loaded into a UART. Hence, data packet is converted into a bit stream using this technique.

The next stage is RS-232/RS-485 line driver, which converts the TTL bit stream into higher voltage bit stream in order to communicate with the CL4790. There are two input signals that are sent to the MCU. There is 12-bit digital measurement signal and 6-bit acknowledgement and select signal. 12-bit digital measurement signal comes from one of the five measurement devices that had been polled by the MCU and the analog to digital converter. 6-bit acknowledgement and select signal come from the receiver. 6-bit acknowledgement and select signal are composed of 3-bit select signal and 3-bit acknowledgement signal. 12-bit measurement signal is sent to the MCU's universal asynchronous receiver/transmitter or UART. 6-bit acknowledgement signal is split between 3 separate 1-bit acknowledgement signals that are sent to an acknowledgement circuit and 3-bit select signal. 12-bit output of the UART and 3-bit select signal are then sent for sampling. The sampling code uses the MCU clock to choose a single 12- bit measurement signal and appends 3-bit label to create 15-bit output that is sent to the RS-232/RS-485 Line Driver and then to the transmitter. The sampling code also creates 3-bit select signal to be sent to MUX for measurement device/sensor selection.

The receiver circuit is controlled by the second microcontroller C8051F340 that is referred as receiver MCU. There are two input signals that are sent to the receiver MCU. These signals are 12-bit measurement signal from the transmitter and 3-bit select signal. The 15-bit measurement signal from the transmitter is composed of 12-bit measurement signal from one of the measurement devices. 3-bit label designates the measurement device. 15-bit signal is first split into two signals, 3-bit label and 12-bit measurement signal. 12-bit signal is then sent via microcontroller's UART to digital to analog converter. MCP 4822 is the device that is selected as D/A converter. Timing Diagram for the DAC conversion process is shown in Fig. 16.5.

After D/A conversion, the measurement signal is sent for signal conditioning. 3-bit label is sent into two different segments of the program. It is first sent to the SC-2345 signal conditioning box using 3-digital input ports for identification of the measurement signal. 3-bit acknowledgement signal is then combined with the 3-bit select signal to form 6-bit signal and sent to the UART at the receiver

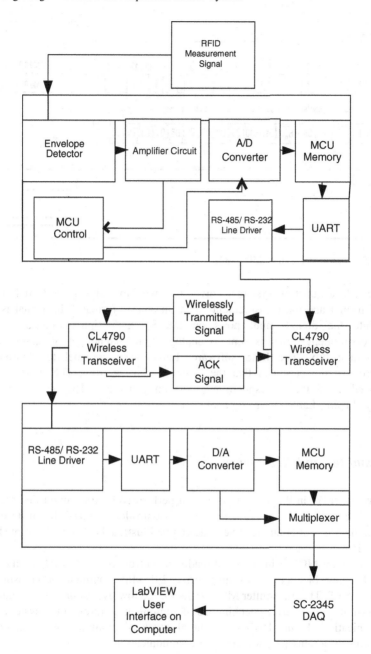

Fig. 16.4 Block diagram of the communication system

microcontroller and then fed to the RS-232/RS-485 Line Driver to be transmitted back to the transmitter to confirm the reception of the measurement signal for the corresponding sensor at the receiver end.

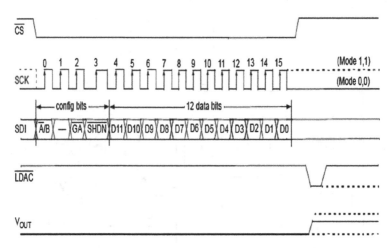

Fig. 16.5 Timing diagram of MPC 4822

The data acquisition system receives the two input signals: a 3-bit Label in digital format and the measurement signal in analog format. 3-bit signal is then sent into the digital input ports on the SC-2345 signal-conditioning box and measurement signal is sent to analog input channel on SC-2345 signal-conditioning box. The DAS software, LabView, uses label to identify the measurement device/sensor that the signal is coming from. The signal is then processed using the corresponding algorithm generated. The results are then displayed using LabView front panel.

Transmitter and Receiver MCU

The critical tasks in the system have been performed by transmitter and receiver MCU. The code that is developed for microcontroller, C8051F340, to do operations at the transmitter and receiver side can be illustrated as block diagram shown in Fig. 16.6.

The receiver MCU is in charge of reading data from the system ADC, amending error-checking bits and performing RS-232/RS-485 communication using the system UART. The transmitter MCU waits for the low byte of data to be sent from the transmitter, does error checking and communication through the system DAC communication. Figure 16.6 is a general block diagram for the receiver and transmitter programs functionality and communication.

The top loop represents the functionality of the transmitter program and the bottom loop represents the receiver functionality. The transmitter MCU handles ADC communication using 5 general I/O pins. The pins are labeled as Strobe, Data, Clock, Chip Select Low and Signal and they use ports 3.0, 3.1, 3.2, 3.3, 3.4. For ADC communication, MCU sends out a serial clock and command signal.

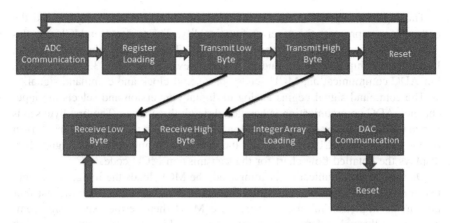

Fig. 16.6 Operational block diagram for the code used for MCU at the receiver and transmitter sides

The command signal begins analog to digital conversion and selects the input channel. ADC communication is handled through three bytes. The first byte sends command signal to the ADC then sends a Strobe signal at the beginning of data out of the ADC. The MCU then records the data into an integer array. Figure 16.7 displays the detailed flow chart for the transmission MCU code.

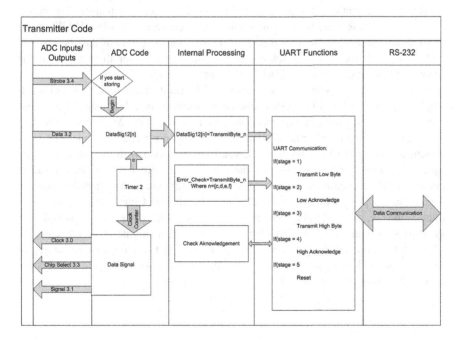

Fig. 16.7   Transmitter MCU flowchart

The top loop represents the functionality of the transmitter program and the bottom loop represents the receiver functionality. The transmitter MCU handles ADC communication using 5 general I/O pins. The pins are labeled as Strobe, Data, Clock, Chip Select Low and Signal and they use ports 3.0, 3.1, 3.2, 3.3, 3.4. For ADC communication, MCU sends out a serial clock and command signal.

The command signal begins analog to digital conversion and selects the input channel. ADC communication is handled through three bytes. The first byte sends command signal to the ADC then sends a Strobe signal at the beginning of data out of the ADC. The MCU then records the data into an integer array. Figure 16.7 displays the detailed flow chart for the transmission MCU code.

Once ADC communication is completed, the MCU loads the integer array into two transmission registers. The value 0×05 is then loaded into the highest four bits of the high transmission register. The MCU then verifies successful communication through acknowledgement signal checking to verify successful high and low byte reception. During MCU/UART communication the transmitter sends the data in two separate byte registers. The transmitter sends the first data byte then waits for an initial acknowledgement value of 0xAA. Once the first acknowledgement is received the second data byte with the amended error-checking bit is sent. The transmitter then waits for a second acknowledgement value of 0xFF. After reception of the second acknowledgement the program then resets back to the original state and begins sampling the ADC again and cycles consistently while there is an active connect to the receiver.

The receiver MCU program is split between three main functions; these are UART communication, internal processing and DAC communication. The receiver executes these functions sequentially and then resets. During UART communication the receiver accepts a data value once the first byte of data is received and sends back 0xAA as an acknowledgement. The receiver then waits for the second byte of data using the UART hardware interrupt.

After the second byte is received the MCU sends back a second acknowledgement value of 0xFF. UART communication is handled using the Timer 1 interrupt. During the internal processing system stage, the MCU loads the data value into an integer array and checks for the correct error checking bits. The most significant four bits of the second received byte is loaded with the error checking value. The error checking bits are then compared with 0x05, if this value is present, DAC communication is began. DAC communication is handled through 4 general I/O pins, which are; Clock, Data, Latch and Chip Select. DAC communication is handled using timer 2 interrupts. Using the clock signal and synchronized data the MCU first sends a 4-bit command signal and then sends the 12-bit data value most significant bit. After sending the complete 16-bit signal the latch pin is lowered momentarily and then raised again to begin conversion. Figure 16.8 is a detailed flowchart of the receiver code.

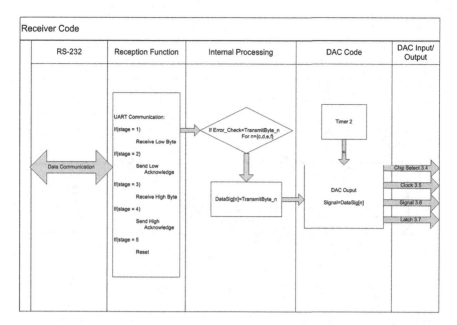

Fig. 16.8   Receiver MCU flowchart

DAQ Software and User Interface

LabVIEW is used as DAS software to communicate with DAQ hardware to specify the measured signal for the corresponding sensor on patient or elder. This is accomplished using the signal from the output of D/A converter and 3-bit label sent by the transmitter. There are analog input, analog output, digital input and digital output modules and dedicated slots in SC-2345 unit for signal conditioning. It is possible to select up to eight modules for analog I/O and eight modules for digital I/O in SC-2345 unit. 3-bit label signal is sent to digital input ports, P0.1—17, P0.2—49, P0.3—47 on the SC-2345 signal conditioning unit. DAQ hardware DaqPad-6020E has the same port numbers, which are used by LabView code that is developed to identify the corresponding sensor. The measurement signal from D/A converter output is sent to analog input channel on SC-2345 signal conditioning unit, which is again polled by LabView code from DAQ hadware. The LabView code performs the voltage to physical variable measurement conversion and displays results to the health-care professional.

The program also has statistics section, which is used to predict the behavior of each measured variable versus time. The flowchart of the LabView code used at the base station is illustrated in Fig. 16.9.

The block diagram showing the operation of LabView code for temperature sensor is illustrated in Fig. 16.10.

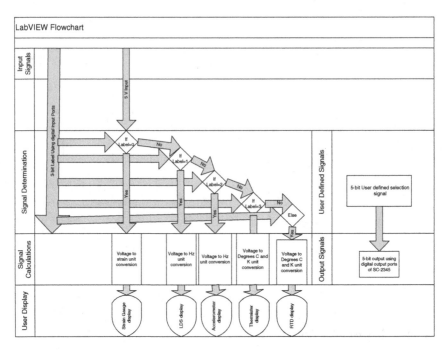

Fig. 16.9 LabVIEW code flowchart

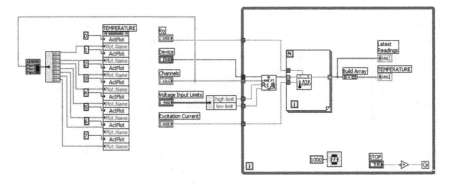

Fig. 16.10 Block diagram of temperature sensor measurement for LabVIEW code

Simulations

The transmitter and the receiver side of the communication system are simulated using National Instruments Multisim V11 circuit simulator. Transmitter side of the circuit has interfacing circuit which performs analog to digital conversion for the sensor output signal that will be fed to stamp microcontroller as discussed earlier. The simulated circuit accomplishing this task is shown in Fig. 16.11. The sensor is

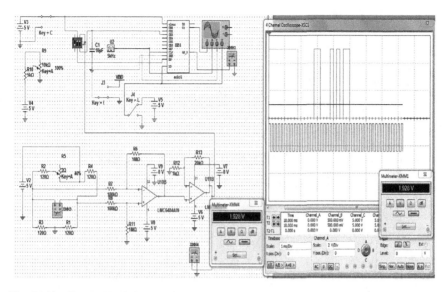

Fig. 16.11   Simulated ADC transmitter interfacing circuit

included in the simulation to measure the response of the sensor located on the patient's body. In the simulation, the sensor produces output of 1.928 V, which is, then send to ADC interfacing circuit. The digital conversion is successfully done using the circuit designed with ADC MAX 186 and digital output signal stream that will be sent microcontroller is illustrated on Fig. 16.11.

Individual sensors that are placed on patient's body are also simulated as stand alone circuits without interfacing circuits. One of the important sensor that is located on patient's body is a temperature sensor, thermistor. This simulation is shown in Fig. 16.12. The temperature that is measured by thermistor is calculated using a third order Steinhart-Hart equation as

$$\frac{1}{T} = A + B\ln(R) + C(\ln(R))^3 \tag{1}$$

where T is temperature in Kelvin, R is the resistance in ohms, A, B, C are Steinhart-Hart coefficients that depend on the type of the semiconductor material. The equation given in (16.1) can be simplified by setting C = 0 and B = 1/b. The equation then simplifies to

$$R = R_0 e^{\beta\left(\frac{1}{T} - \frac{1}{T_0}\right)} \tag{2}$$

In Eq. (16.2), R_0 is defined as base resistance T_0 is defined as base temperature. Beta is calculated from

$$\beta = \ln\left(\frac{R}{R_0}\right)\left(\frac{T_0 T}{T_0 - T}\right) \tag{3}$$

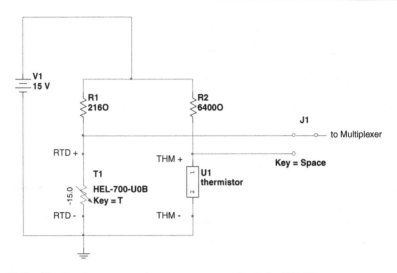

Fig. 16.12 Simulated temperature input measurement circuit for WDAS

Taking the natural log of Eq. (16.3) gives the following equation

$$\ln(R) = \beta\left(\frac{1}{T} - \frac{1}{T_0}\right) + \ln(R_0) \qquad (4)$$

Plotting $\ln(R)$ versus $(1/T - 1/T_0)$ gives a linear response with the gradient β. The schematic capture of the transmitter circuit with interfacing components that is used to obtain board layout is illustrated in Fig. 16.13. This circuit is also used to get the layout of the final board that is routed. The PCB board layout for the schematic in Fig. 16.13 is obtained using Ultiboard as illustrated in Fig. 16.14 below.

Experimental Results

The proposed general purpose, long range and low cost multiple input multiple output (MIMO) wireless data acquisition sensor system for health-care applications is implemented and tested. The breadboard and the final routed board for the transmitter circuit with interfacing components that exists on the second stage are shown in Fig. 16.15.

The test has been performed in outdoor and indoor environments to measure 1 Hz sinusoidal signal and temperature signal using thermistor as a temperature sensor. 1 Hz AC signal is used to see the measurement accuracy in the case when sensor output varies. The transmitted signal measured on oscilloscope is shown in Fig. 16.16.

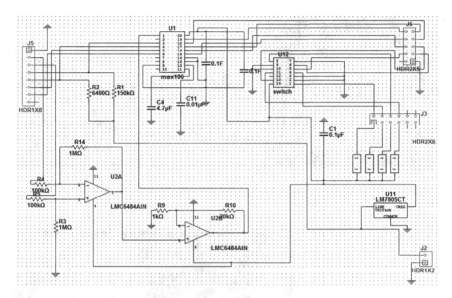

Fig. 16.13 Schematic capture of the transmitter circuit

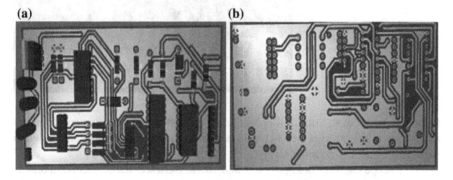

Fig. 16.14 Transmitter board layout with interfacing components. **a** Top board view, **b** Bottom board view

The specific measurement results in Figs. 16.17, 16.18 and 16.19 are the measurement results for the AC signal transmitted using wireless data acquisition system. The transmitter and receiver have been separated up to 20 mile distance during outdoor testing. The base station where receiver and DAQ system with all the associated hardware and software are located inside a closed structure where power was available. The transmitter is located at a location with a portable power supply module. It has been observed that the transmission and reception was good when there were no obstacles such as trees and buildings between the receiver and transmitter. Better communication was possible with the replacement of longer-range receiver and transmitter antenna system. The measurement results for the

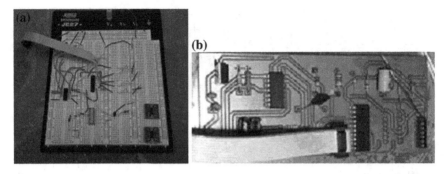

Fig. 16.15 Constructed transmitter board with interfacing components. **a** Breadboard, **b** final board

Fig. 16.16 Transmitted
1 Hz analog signal

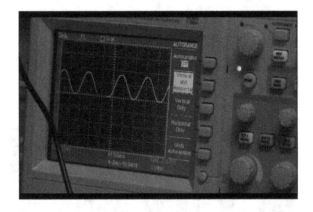

measured signal at the transmitter side after it is processed and converted digital format and at the receiver side before it is converted to analog format are shown in Figs. 16.17 and 16.18, respectively.

In the scope captures in Figs. 16.17 and 16.18, the trace that is colored in blue is the measurement signal. The signal that is colored in orange is the clock signal.

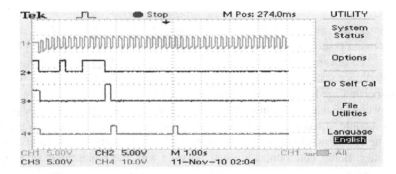

Fig. 16.17 Measurement result at the receiver side of WDAS

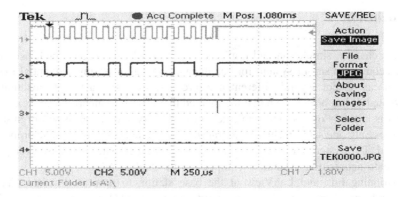

Fig. 16.18 Measurement result at the transmitter side of WDAS

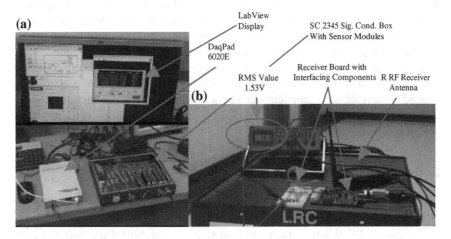

Fig. 16.19 **a** LabView setup with user display; **b** Receiver circuit with interfacing components and RF receiver antenna

The LabView setup with the display of the front panel for the measured physical variable is illustrated in Fig. 16.19a. Figure 16.19b shows the set up for the receiver circuit including RF receiver antenna, receiver circuit and interfacing components.

LabView front panel in Fig. 16.19 illustrates the received 1 Hz signal, which is identical to the transmitted signal from 20 miles away using the established WDAS. In addition, Figs. 16.17 and 16.18 confirm the successful communication using the receiver and transmitter circuit with their interfacing components. As a result, it has been illustrated that WDAQ system that has been implemented can be used to measure, transmit and receive sensor signals up within 20 mile range.

Conclusion

In this chapter, simulation and implementation of a long-range portable wireless data acquisition sensor system for health care applications using sequential logic, which can accept several different measurement signals at a remote location is presented. It has been shown that the method that is used is superior to the existing systems in terms of cost and implementation. This type of wireless sensor system requires only one set of receiver and transmitter antenna system using the proposed method. The system has a longer communication range with RF transceiver pair. The data acquisition system is equipped with DaqPad 6020E, SC 2345 signal conditioning box and LabView at the base station to process the data that is received and display the final measured values. WDAQ sensor system is also capable of doing statistical analysis and can be used as a warning system to raise the flag when such as condition occurs. Simulated and measured values are presented and successful communication using the proposed system is illustrated in an outdoor environment within 20 miles of range. The communication range can be increased using longer range RF transceiver. This system gives real-time remote monitoring ability of patients and elders and as a result plays vital role for home care facilities and health care institutes. The system that is developed can be utilized in several other applications including structural health monitoring systems, and security systems.

References

1. Namjun Cho, Seong-Jun Song, Sunyoung Kim, Shiho Kim and Hoi-Jun Yoo, "A 8-uW, 0.3-mm^2 RF-Powered Transponder with Temperature Sensor for Wireless Environmental Monitoring," ISCAS 2005. IEEE International Symposium, pp. 4763–4766, 23–26 May 2005.
2. G. Y. Daniloff, "Continuous glucose monitoring: Long-term implantable sensor approach," *Diabetes Technol. Ther.*, vol. 1, pp. 261–266, 1999.
3. J. W. Mo and W. Smart, "Lactate biosensors for continuous monitoring," *Front Biosci.*, vol. 9, pp. 3384–3391, 2004.
4. K. V. Dmytruk, O. V. Smutok, O. B. Ryabova, G. Z. Gayda, V. A. Sibirny, W. Schuhmann, M. V. Gonchar, and A. A. Sibirny, "Isolationand characterization of mutated alcohol oxidases from the yeast Hansenula polymorpha with decreased affinity toward substrates and their use as selective elements of an amperometric biosensor," *BMC Biotechnol.*, vol. 7, p. 33, 2007.
5. H.Norton, *Handbook of Transducers*, Englewood Cliffs, NJ: Prentice Hall, 1989, Ch. 2.
6. N. Kurata, B.F. Spencer Jr, M.R. Sandoval, "Risk Monitoring of buildings with wireless sensor networks," Structural Control and Health Monitoring, vol. 12, pp. 315–327, 2005.
7. J. Beutel, M. Dyer, M. Hinz, L. Meier and M. Ringwald, "Nextgeneration prototyping of sensor networks," in SenSys'04: Proceedings of the 2nd international conference on Embedded networked sensor systems. New York, NY, USA: ACM, 2004, pp. 291–292.
8. F. Schmidt and G. Scholl, "Wireless SAW Identification and Sensor Systems," International Journal of High Speed Electronics and Systems, vol. 10, pp. 1143–1191, 2000.
9. M. Ferrari, V.F.D. Marioli and A. Taroni, "Autonomous Sensor Module with Piezoelectric Power Harvesting and RF Transmission of Measurement Signals," Instrumentation and Measurement Technology Conference, pp. 1663–1667, 2006.

10. P. Mitcheson, E.Y.G. Rao, A. Holmes and T. Green, "Energy Harvesting From Human and Machine Motion for Wireless Electronic Devices," Proceedings of the IEEE, vol. 96 No. 9, pp. 1457–1486, 2006.
11. Klaus Finkenzeller, *RFID Handbook*, Wiley, 2nds Edition, April 2003.
12. M. Bhattacharya, C.H. Chu and T. Mullen, "A Comparative Analysis of RFID Adoption in Retail and Manufacturing Sectors," IEEE International Conference on RFID, pp. 241–249, 2008.
13. D.A. Rodriguez-Silva, F.J. Gonzalez-Castano, S. Costas-Rodriguez, J.C. Burguillo-Rial, R. Gentile, S. Stanca and R. Arona, "Quantitative assessment of the benefits of RFID technology for libraries: a trans- European study," Automatic Identification Advanced Technologies, 2007 IEEE Workshop on, pp. 128 –133, June 2007.
14. S. Sarma, "Towards the 5c tag," White Paper, MIT Auto ID Center, 2001.
15. EPC Global, "Class 1 Generation 2 UHF Air Interface Protocol Standard," http://www.epcglobalinc.org/standards/uhfc1g2.
16. A. Juels, "RFID security and privacy: a research survey," Selected Areas in Communications, IEEE Journal on, vol. 24, no. 2, pp. 381–394, Feb. 2006.
17. CL4790-1000-232, ConnexLink-Wireless Cable Replacement System, Datasheet.

Chapter 17
Performance Improvement of RFID Systems

Abdullah Eroglu

Introduction

Radio frequency identification devices (RFIDs) have been used widely in industrial and medical applications due to the fact they bridge the real and the virtual worlds and enable information transfer at a large scale in a cost effective way. These devices use radio waves for non-contact reading and are effective in manufacturing and several other applications where bar code labels could not survive.

An RFID system includes mainly three components: *a tag or transponder* located on the object to be identified, *an interrogator* (reader) which may be a read or write/read device, and *an antenna* that emits radio signals to activate the tag and read/write data to it as shown in Fig. 17.1.

RFID tags are classified as passive, semi-active (or semi-passive), and active. Passive tag operates on its own with no source and makes use of the incoming radio waves as energy source. Active tags use the battery power for continuous operation. Semi-active tags use battery power for some functions but utilize the radio waves of the reader as an energy source for its own transmission just like passive tags. Passive tags are very popular due to the fact that they operate without need of external power source. One of the biggest challenges of passive RFID tags is their communication distance. The operating range of a RFID system is based on tag parameters such as, tag antenna gain and radar cross section, distances between readers, operating frequency, transmission power from reader to the tag, and gain of the reader antenna. As a result, tag antenna performance plays very important role in identification of the communication distance with operational frequency. Typical operational frequencies and their corresponding communication distances

A. Eroglu (✉)
Indiana University–Purdue University Fort Wayne, 2101 E. Coliseum Blvd, Fort Wayne, IN 46805, USA
e-mail: eroglua@ipfw.edu

S. C. Suh et al. (eds.), *Applied Cyber-Physical Systems*,
DOI: 10.1007/978-1-4614-7336-7_17,
© Springer Science+Business Media New York 2014

Fig. 17.1 Basic diagram of
RFID system

for RFID tags are given in Table 17.1. As seen from Table 17.1, the communication distance for ultra high frequency range (UHF) is much higher than other conventional RFID frequency bands [1].

Planar type antennas such as patch antennas are widely used in RFID systems due to their several advantages including low profile, light weight, easy fabrication and conformability to mounting hosts in addition to size reduction and bandwidth enhancement. The advantages and disadvantages of patch antennas in RF systems are shown in Table 17.2. One of the cost effective solution to this challenge is the use of higher gain patch antennas with low profile. The recent advancements in EBG structures in antenna designs show that they reduce the propagating surface waves. Reduction in surface wave improve the antenna gain, directivity and bandwidth [2, 3]. In addition, the profile of the patch antenna can be further lowered when high permittivity material is used.

In this chapter, performing patch antenna design with EBG structure and high permittivity material is given for UHF RFID systems. The proposed antenna is designed and simulated with 3D electromagnetic simulator, Ansoft HFSS. Several important antenna parameters have been investigated for performance improvement and simulated results are presented. Results of this paper can be used in patch antenna design to improve its performance for applications such as biomedical and asset-tracking.

Table 17.1 RFID tags and communication distance

RFID frequency band	Frequency band	Typical communication distance (m)	Common application
125–134.2 kHz and 140–148.5 kHz	Low frequency	Up to ~½	Animal tracking access control product authentication
13.553–13.567 MHz	High frequency	Up to ~1	Smart cards shelve item tracking airline baggage maintenance
858–938 MHz, 902–928 MHz, North America	Ultra high frequency	Up to 10	Pallet tracking carton tracking electronic toll collection parking lot access
2.45 GHz/5.8 GHz	Microwave	Up to 2	Electronic toll collection airline baggage

Table 17.2 Advantages and disadvantages of patch antennas

Advantages	Disadvantages
Light weight	Narrow bandwidth
Low volume	Not the greatest gain
Low profile	Incapable of handling high power
Conformal to multiple surfaces	Extraneous radiation from junctions and feeds
Low fabrication cost	Excitation of surface waves
Feeding and matching can be fabricated simultaneously with the antenna structure	Poor efficiency from use of material with a high dielectric constant

Patch Antenna Design for RFID Systems

Conventional patch antenna shown in Fig. 17.2 is designed based on the required resonant frequency (fr), size, and bandwidth. The size and bandwidth requirements determine the dielectric constant (εr) and height (h) of the substrate. Increasing the height of the substrate increases the bandwidth, but it also increases the size of the antenna and could increase propagation of surface waves, which causes performance degradation. Meanwhile, increasing the dielectric constant decreases the size of the antenna but narrows the bandwidth. Therefore, the substrate's dielectric constant and height must be selected carefully depending on the application. Once a dielectric substrate is selected, the width (W) and length (L) of the radiating patch can be calculated. The width of the patch antenna is calculated from [4, 5]

$$W = \frac{c}{2f_r}\left(\frac{2}{\varepsilon_r + 1}\right)^{1/2} \tag{17.1}$$

The length of the patch is found using

$$L = \frac{c}{2f_r\sqrt{\varepsilon_{eff}}} - 2\Delta L \tag{17.2}$$

where

Fig. 17.2 Patch antenna: **a** *top view*, **b** *side view*

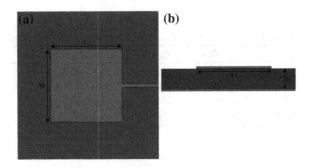

$$\Delta L = 0.412h \frac{\left(\varepsilon_{eff} + 0.3\right)\left(\frac{W}{h} + 0.264\right)}{\left(\varepsilon_{eff} - 0.258\right)\left(\frac{W}{h} + 0.8\right)} \tag{17.3}$$

and

$$\varepsilon_{eff} = \frac{\varepsilon_r + 1}{2} + \frac{\varepsilon_r - 1}{2}\left(1 + \frac{12h}{W}\right)^{-1/2} \tag{17.4}$$

Square patch antenna is designed by using Eqs. (17.2–17.4) with an Nelder–Mead optimization technique as

$$l = \frac{c}{2f_r\sqrt{\varepsilon_{eff}}} - 2(0.412h)\frac{\left(\varepsilon_{eff} + 0.3\right)\left(\frac{l}{h} + 0.264\right)}{\left(\varepsilon_{eff} - 0.258\right)\left(\frac{l}{h} + 0.8\right)} \tag{17.5}$$

To obtain a desirable return loss at the resonant frequency, a microstrip patch antenna must be matched to the transmission line feeding it. Before employing any matching technique, the resonant input impedance must be calculated. The transmission line model and cavity model can be applied to calculate the input impedance at the edge of the patch antenna. Once the edge input impedance is calculated, the different matching techniques can be employed. The transmission line model and cavity model with and without mutual effects included for various dielectric permittivity constants and dielectric substrate thicknesses have been compared through simulations with HFSS are shown in Tables 17.3 and 17.4. The transmission line model with mutual effects included was found to be the best estimate of the edge impedance for the resonant frequency of interest (915 MHz). These results were tested with probe-fed and inset-fed patch simulations using the following inset equation

$$Z_{in}(y = y_0) = Z_{in}(y = 0)\cos^2\left(\frac{\pi y_0}{L}\right) \tag{17.6}$$

It has been found that probe-fed antenna gives better performance than the inset-fed antenna.

Table 17.3 $Z_{in}(\Omega)$ of patch with $h = 1.5$ mm and $f_r = 915$ MHz

ε_r	Z_{in} (TL)	Z_{in} (TL–M)	Z_{in} (Cavity)	Z_{in} (Cav–M)	Sim (Z_{in})
6.5	232.4	186.4	678.0	394.1	117.6
8.9	267.0	216.0	894.0	499.2	158.5
9.8	278.9	226.5	975.0	539.0	172.8
12.88	316.1	260.1	1252.2	675.9	210.7

Table 17.4 $Z_{in}(\Omega)$ of patch with $\varepsilon_r = 9.8$ and $f_r = 915$ MHz

h (mm)	Z_{in} (T L)	Z_{in} (TL–M)	Z_{in} (Cavity)	Z_{in} (Cav–M)	Sim (Z_{in})
1.5	278.9	226.5	975.0	539.0	172.8
2.0	278.9	226.5	975.0	538.9	169.7
2.5	278.9	226.4	975.0	538.7	171.2
3.0	278.9	226.4	975.0	538.5	171.3
3.5	278.9	226.4	975.0	538.2	168.6

Patch Antenna Design with EBG Structures

EBG structure can be approximated by lumped LC elements when the operating wavelength is large compared to the periodicity as shown in Figs. 17.3 and 17.4. The small gaps between the patches generate a capacitance and the current along adjacent patches produces an inductance [6]. The impedance of a parallel resonant LC circuit is given by

$$Z = \frac{j\omega L}{1 - \omega^2 LC} \tag{17.7}$$

where the capacitance and inductance relations are obtained using

$$C = \frac{W\varepsilon_0(1 + \varepsilon_r)}{\pi} \cosh^{-1}\left(\frac{W + g}{g}\right) \tag{17.8}$$

$$L = \mu_0\mu_r h \tag{17.9}$$

The resonant frequency is then found from

$$f_r = \frac{1}{2\pi\sqrt{LC}} \tag{17.10}$$

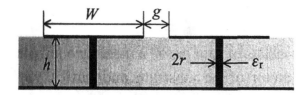

Fig. 17.3 EBG structure implementation

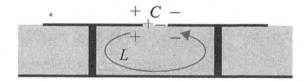

Fig. 17.4 LC equivalent model of EBG structure

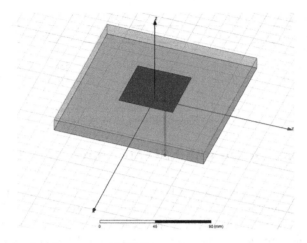

Fig. 17.5 Simulated patch antenna with EBG structure

The approximate size that is allowed for the EBG structure is obtained through the relation given below.

$$BW \propto \frac{L}{C} \qquad (17.11)$$

For a frequency band gap centered at 915 MHz (UHF RFID), a patch antenna surrounded by a mushroom-like EBG structure on a typical substrate would require gaps that are small.

Patch antennas that are modeled and simulated with 3D electromagnetic simulator, HFSS, given in Fig. 17.5. Figure 17.6 show the simulated patch antenna with and without EBG structure.

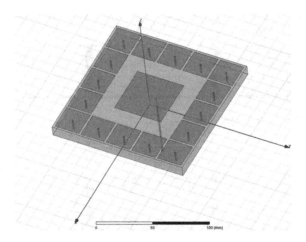

Fig. 17.6 Simulated patch antenna with EBG structure

Simulation Results

The performances of the patch antenna with and without an EBG structures are compared when high permittivity material, Alumina, with dielectric constant $\varepsilon_r = 9.8$ and thickness, h = 10 mm is used as a substrate. The performance comparison results for the patch antenna with and without EBG structures are shown in Figs. 17.5 and 17.6 have been done using 3D electromagnetic simulator HFSS. Table 17.5 outlines the physical dimensions used during the simulation of both configurations.

The gain of the patch antenna on xz and yz planes with and without EBG structures are shown in Figs. 17.7 and 17.8, respectively.

Table 17.5 Physical dimensions of the patch antennas simulated

Parameter	Patch antenna without EBG	Patch antenna with EBG	EBG structure	
Substrate			Width	23.6 mm
Height	10 mm	10 mm	Gap	2.01 mm
Dielectric constant	10.2	10.2	Radius of vias	1 mm
Length	130 mm	130 mm	Distance from patch	16.17 mm
Width	130 mm	130 mm	Number of units	16
Patch			Number of rows	1
Length	46.65 mm	46.5 mm		
Width	46.65 mm	46.5 mm		
Feed location				
y_o	14.925 mm	16.4 mm		

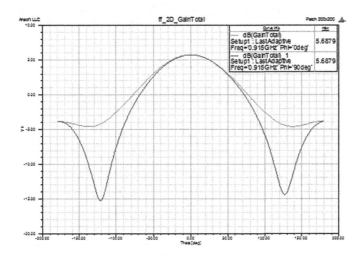

Fig. 17.7 Gain of patch antenna without EBG structures in rectangular coordinate system

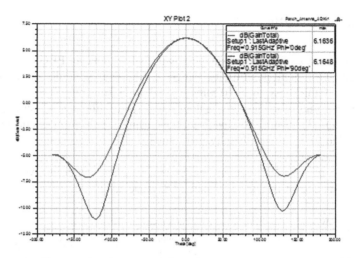

Fig. 17.8 Gain of patch antenna with EBG structures in rectangular coordinate system

As seen from the gain patterns in rectangular coordinate system, the gain improvement is around 0.5 dB when EBG structure is implemented. The strength of the minor lobe is also reduced with the use of EBG structures. This can be better demonstrated with the radiation pattern in polar coordinate system as shown in Figs. 17.9 and 17.10. Figure 17.9 shows that radiation pattern without EBG structure whereas Fig. 17.10 gives the radiation pattern of the patch antenna with EBG structure.

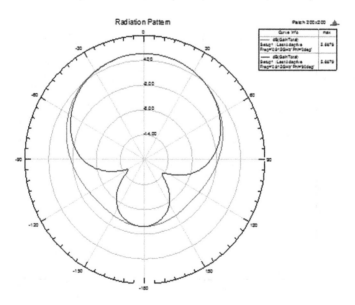

Fig. 17.9 Gain of patch antenna without EBG structures in polar coordinate system

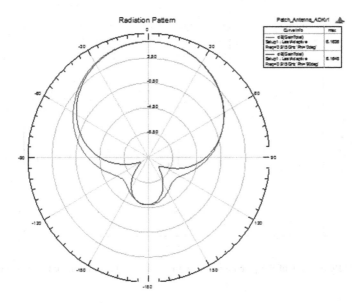

Fig. 17.10 Gain of patch antenna with EBG structures in polar coordinate system

The return losses of the patch antenna with and without EBG structure are given in Figs. 17.11 and 17.12.

The return loss of the patch antenna seems to be unaffected with the implementation of EBG structures. The bandwidth of the antenna with EBG structure is narrower. Front to back ratio of the antenna with EBG structure is around 1.5 dB higher than the same antenna without EBGs. The results showing the performance

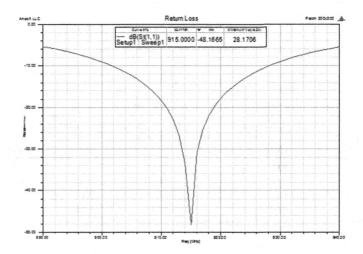

Fig. 17.11 Return loss of the patch antenna without EBG structures in polar coordinate system

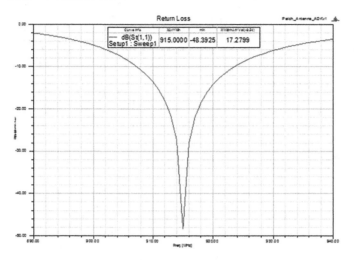

Fig. 17.12 Return loss of the patch antenna with EBG structures in polar coordinate system

of the antenna are tabulated in Table 17.6. It has been observed that the size of the ground plane also affects the radiation characteristics of the antenna. The performance of the antenna with different ground planes are also tabulated and given in Tables 17.7 and 17.8.

Table 17.6 Performance results of patch antenna with and without EBG structures

	Resonant frequency (MHz)	Return loss (dB)	Bandwidth (MHz)	Peak gain (dB)	Front-to-back (dB)
Patch without EBG	915	−48.17	28.17	5.6879	9.5497
Patch with EBG	915	−48.39	17.28	6.1648	11.0656

Table 17.7 Performance results of patch antenna without EBG structures for different ground plane sizes

Ground plane (mm × mm)	Resonant frequency (MHz)	Return loss (MHz)	Bandwidth (MHz)	Peak gain (dB)
130 × 130	915	−48.39	17.28	6.1648
200 × 200	919	−18.33	14.46	6.1917
300 × 300	918	−14.79	12.14	5.6481
500 × 500	918	−15.69	12.84	5.8219

Table 17.8 Performance results of patch antenna with EBG structures for different ground plane sizes

Ground plane (mm × mm)	Resonant frequency (MHz)	Return loss (MHz)	Bandwidth (MHz)	Peak gain (dB)
130 × 130	915	−48.39	17.28	6.1648
200 × 200	919	−18.33	14.46	6.1917
300 × 300	918	−14.79	12.14	5.6481
500 × 500	918	−15.69	12.84	5.8219

Conclusion

In this chapter, design and simulation of better performing patch antenna for UHF RFID systems with low cost is introduced using EBG structures. The simulated results for the patch antenna with and without EBG structures have been compared for antenna performance at the frequency of interest. It has been shown that gain of the antenna improve with the proposed design method. The improvement over bandwidth and return loss have also been observed with different conditions when EBG structure is implemented. The manufacturing process of the antenna is also quite straightforward since EBG structures are integrated to the overall structure. Patch antenna with EBG structure presented in this paper can be used in several applications where performance is an important parameter. This includes biomedical and asset-tracking applications.

References

1. Omni-ID® Ultra. "*Omni-ID Passive RFID Tags: High Performance Radio Frequency Identification for RFID Asset Tracking*". Omni-ID, 2011. Web. 22 Dec. 2011.
2. T. T. Nguyen, D. Kim, S. Kim, J. Jang, "Design of a wideband mushroom-like electromagnetic bandgap structure with magneto-dielectric substrate," 6th International Conference on Information Technology and Applications, Nov. 2009, pp. 130–135.
3. Tan, M. N. Md.; Rahman, T. A.; Rahim, S. K. A.; Ali, M. T.; Jamlos, M. F., "Antenna array enhancement using mushroom-like electromagnetic band gap (EBG)," Antennas and Propagation (EuCAP), 2010 Proceedings of the Fourth European Conference on, vol., no., pp. 1–5, 12–16 April 2010.
4. Balanis, C. A., Antenna Theory: Analysis and Design (3rd edition), John Wiley & Sons, New York.
5. Fang, D. G., Antenna Theory and Microstrip Antennas, CRC Press Taylor & Francis Group, Boca Raton.
6. Yang, Fan, and Yahya Rahmat-Samii. *Electromagnetic Band Gap Structures in Antenna Engineering*. Cambridge, UK: Cambridge UP, 2009.

Chapter 18
Thinking Embedded, Designing Cyber-Physical: Is it Possible?

Deniz Koçak

Introduction

Computers are getting smaller and their capabilities increase incredibly. However, even today computing includes human control to enter data and interpret the results. Even for modern systems like transportation, healthcare and similar safety critical missions require human control to ensure the safety of human life. As a result of the increase in complexity, it has been getting harder to manage these systems just by humans for every possible case. For example railway signalization is already very complex and hard task to handle just by human capabilities. Huge number of variables in such an equation needs intelligent systems which can also control physical world in addition to data acquisition and report. Conventional control systems programmed for specific tasks, are unable to adapt themselves to changing conditions and environments. Therefore less control on physical world conditions due to this insufficient intelligence capacity makes them unusable.

Adaptation of large scale systems into the physical world is a great achievement, but an unstructured problem and includes many items from different domains.

Cyber-Physical Systems

Cyber-physical systems stay in the border of cyber and physical worlds, which requires more collaboration between these two worlds than in the past. This collaboration depends on computing, communication and control capabilities of deployed systems. In Fig. 18.1, data flow from physical world to the cyber world depends on the abilities of various kinds of sensors connected to the physical

D. Koçak (✉)
BITES, METU-Technopolis, Ankara, Turkey
e-mail: deniz.kocak@linux.org.tr

S. C. Suh et al. (eds.), *Applied Cyber-Physical Systems*,
DOI: 10.1007/978-1-4614-7336-7_18,
© Springer Science+Business Media New York 2014

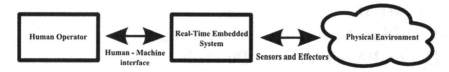

Fig. 18.1 Interaction between physical and cyber world

environment. Real-time embedded systems can communicate with humans and other systems in the network. Communication between the machines may be autonomous in order to achieve high availability and responsiveness. Human operator and other nodes in the network may work homogeneously to make right and fast decisions after the analysis of incoming data from the real world within current context. For example, a railway signaling system may decrease the speed of train with respect to the current location and status of the railway without getting a direct command from the operator. Moreover, a train can share this information with other trains closest to it and also with the control center. All this computing and communication processes can take place within a few seconds, which is practically impossible for humans without the help of autonomous systems. All this may happen in a very short time for human reflex, and saves lives of many people.

The transition from conventional embedded real-time devices to cyber-physical systems takes place quite fast. Simple and also single node, task oriented devices mutating into context-aware, multitasking and interactive devices in a network of nodes. Although many concepts like real-time systems, distributed and pervasive computing, sensor networks forms the basis for cyber-physical systems, none of these concepts is enough separately for the complexity of the problems we have to solve. The necessity of interaction with the physical environment, not only for collecting data, but also controlling the variables of its physical environment is more important than in the past (Fig. 18.2). In addition to that, scale of the deployed systems and increasing amount of data creates many design problems that are almost impossible to solve by using conventional methods used today.

Fig. 18.2 Basic tasks of cyber-physical systems

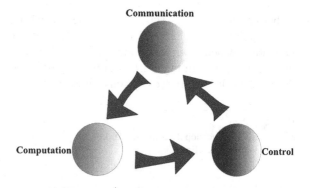

In addition to the complex problems we have been trying to solve, capacity and complexity of embedded real-time systems has been also increasing in parallel to the developments in the hardware technologies. Manufacturers achieved more processing power, memory and permanent storage for embedded systems which are smaller and cheaper than before. Introduction of new networking technologies and human–computer interaction options like LCD screens or even modern touch screens used today increased the interaction and visibility of embedded systems. Increasing number of deployed processing units and sensors increase the complexity of these systems. Even for well understood problems, increasing communication requirements create new problems during the integration of larger and distributed systems.

Design Challenges

Today embedded real-time systems forms a basis for cyber-physical systems which can be seen as the next step of embedded computing evolution. Real-time systems have been an important subject of both academia and companies. However, these systems have recently become prevalent in our daily lives. Many mechanical control units are being replaced by electronic equivalents. Sometimes we may not be aware of this fact, but we will hear these terms frequently in the near future.

Real Time systems can be thought as a system with timing constraints. Timing constraints can change widely from due to safety requirements of the systems. It is very natural for a home theater system not having strict constraints like an airplane or even a car. The start-up time of an entertainment system does not make sense compared to the response time of a car break system. Sometimes real-time computing may be confused with fast computing which is not a true definition. Obviously speed is an important key element in gathering, computing data and reporting the results, however real-time systems are more interested in predictability. In that sense real-time systems can be classified as soft and hard real-time systems.

Main difference between soft and hard real-time systems is the toleration for timing failures. If a system is not tolerable to timing failures and the result is catastrophic, it is in the category of hard real-time systems. On the contrary, if it is tolerable up to a point and result is mostly not catastrophic then it is a soft real-time system [1]. However, this tolerance should be still in the range of limited time which is mostly driven by the requirements of a system. For example 500 ms may be an acceptable delay during the operation of a home theater system, but what if this delay happens in the safety system of a nuclear power plant.

As stated previously, timing constraints are one of the fundamental parts of not only complex cyber-physical systems, but also even part of embedded devices in daily use. However, due to the different requirements of different fields, tolerance to timing failures widely changes between each system. Especially in safety critical systems these constraints are much harder to define, implement and test. Timing constraints are serious problems not only during implementation but also design phase. For even evolutionary software development methodologies [2] there may be dead ends due to insufficient analysis of requirements and design to see whether these requirements will be able to be satisfied during the implementation phase.

Because embedded software systems are closely related to the underlying hardware, sometimes it may be quite hard to make abstractions during the design. Traditionally, portability is not a main concern for embedded software development. System designers tend to stick on a hardware platform that they know and also they have tested previously, during the design of a new system. However, new developments and increasing variety in hardware platforms force designers and developers spend more time on portability issues. Mobile platforms have been evaluating for years and today the boundaries between mobile and conventional computing platforms like personal computers are becoming blurry. Despite these great achievements in hardware technology, the origin of these products still depends on semi-conductor technologies which make them vulnerable to the common weakness of semi-conductor technologies. More data requires more storage space, but without inventing new approaches or even sometimes without shifting the paradigm of designing new hardware systems, this can be achieved by multiplying the number of transistors in a hardware system.

In many cases the process of computing in an embedded system includes gathering data and doing computations by using this data with respect to the logic defined earlier. Data would be anything like temperature of a warehouse or distance of an obstacle from a car which are current data from the physical world. Business logic inside he embedded system may start the cooler in the warehouse or warn the driver about the proximity of an obstacle on the way. The interaction between the physical world and business logic inside the system compose the heart of a cyber-physical system. Reaction of the system also done by the system automatically as in the cooler system of a warehouse or can be delegated to another system or even to a human like the driver of car. These are basic and common systems in our daily lives. More sophisticated system exists like autopilot systems and traffic control systems. However, as mentioned previously all of them are vulnerable because of the common weakness of hardware platforms.

Worst Case Execution Time

Worst case execution time analysis is an important aspect for the design of a system in terms of timing predictability and safety. In the development of safety–critical systems or even simpler systems, timing analysis can tell us the capability of our software in combination with the underlying hardware platform. Analysis is directly related with the design to ensure that design meets requirements. In hard or soft real-time systems, meeting the requirements of the system is an important

Fig. 18.3 Abstraction layers
between application and
hardware

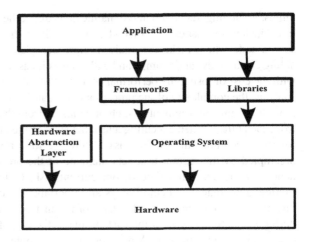

achievement during the lifecycle of the project. It comes into prominence when
developing a safety–critical system or it is hard to adapt one of the agile software
development methodologies for project management. The requirements should be
well understood and the design should meet the requirements. Moreover, rapid
development in the hardware platforms, increasing processing power and related
technologies cause more complexity and also requires adaptation to these new
products. In addition to that improving software development techniques like
component based development, increasing use of commercial off-the-shelf
(COTS) software, distributed computing and even simple software development
libraries widely used today's modern systems.

As shown in Fig. 18.3, every layer between the application and hardware
decrease the predictability. Furthermore, most of these layers are commercial off-
the-shelf products that we need to test and verify with respect to our requirements.

In order to enable ease in scheduling many analysis methods have been widely
used, especially formal methods are still an active research topic of computer
science. However, this paper focuses on advantages and disadvantages of some
common methods and effect of these methods to design process of systems. In
simple terms timing analysis, can be divided into two parts. These are static
analysis methods and dynamic—measurement based—methods.

Static Analysis Methods

Static analysis methods depend on the analysis of source code or machine code
without executing it on a real hardware or simulator [3]. It may use a model of the
underlying hardware architecture which becomes too hard when the underlying
hardware platform utilizes components such as caches, pipelines and branch
prediction [4]. Moreover, basic building blocks of programming like dynamic

jumps, branching, recursion and dynamic calls affect the liability of static analysis. Even in few thousands line of code, almost all of these constructs exists in the source code which make the analysis more complex and harder to handle. Disabling caching, dynamic jumps and calls makes it easier to model the software, but with the cost of losing the performance and flexibility. In the construction of static analysis, the complexity of the processors also cause context dependency problems in multitasking environments which prevents us from thinking the tasks separately from each other. When executing a task the state of the processor is a critical factor for the execution time of this task. Therefore the previous task which may be preempted by the current task sets the state of the processor and cause a dependency for the execution time of our current task. Order of the tasks during the runtime and context switching plays an important role in this case. As a result of this problem counting the clock cycles of the instructions may give wrong results during the static analysis of the system due to the complexity of this multitasking environment and processor. Moreover, frequent context switching between tasks increases complexity much more than we can handle easily. Here, the problem comes with a very simple, but effective solution like reducing the number of tasks or changing the scheduling algorithm used in the system. Although, simplicity saves our life, it may be unavoidable to go further in order to meet the requirements. Hopefully, many techniques and tools exist today for analyzing systems [3], but these will not be covered here.

Another difficulty is the human factor in the design phase. Although, static analysis tools exists today, still analysis techniques requires human mind. In the industry, the modeling of system which is also the base of static analysis, has not been well understood and done successfully by the system engineers. Most of the time system engineers prefer measurement based testing due to its applicability to projects. The lack of knowledge about the platform used in the project or lack of educated man power on the topic ends with a bad reputation of static analysis in industry. Overall, static analysis tends to give good results for simple processors, but with the rapid development in processor technology and increasing complexity it is still an academic research field.

Dynamic Analysis Methods

Dynamic analysis methods depend on measurement of the code running end-to-end on real hardware or simulator [3]. Actual runtime of a code also gives estimation about the timing analysis of the running code. Moreover, it is possible to test each processor and compare the results if they meet the requirements. Because dynamic analysis methods seem to be easy for designers and developers, it is a common practice and widely used in projects. Even further, the results of an analysis for previous projects may be directly used for the new projects when the processor remains same even if a different one from the family is used. This approach may give an idea about the average execution time of a code; however it is not reliable in terms of worst case execution time. Most dominant problem in these techniques is the determination of all path analysis [5]. Determining the set of all possible paths, including branch and loop analysis increase the reliability of this kind of analysis. However, conditionals, loops, recursion, dynamic calls and many other elements widely used today's programming practices, cause large set of path combinations which are practically impossible to test one by one. Also it is a tedious task to determine the paths which gives worst execution time for the system under test [3]. Some paths may suffer from complexity of the processor under test like caches, pipelines which is good to see on a real processor, it is hard to achieve high degree of code coverage. For simple processors or in a limited environment, results may give an idea for the estimation of worst case execution time, but it is still time a consuming and unreliable way of design validation.

Timing analysis may be done separately on different piece of code or bigger components of the system. Decomposing the whole system and piecewise design of the whole system is a common an also useful practice in order to simplify the design process in large projects. However, compositionality of the system should be concerned and when these parts come together the results of the analysis should be preserved after the composition.

Software can be explained as components and the interface between these components in the whole system. Component term here refers different parts of a system, not directly related with the Component Base Software Engineering concept. These can be a set of methods in a library or a remote service which may or not be physically separated. During the design phase each part may be designed separately and then combined into the end system. Compositionality of a system ensures that the characteristic of the whole system can be determined by the combination of each component and the rules used to combine these parts [6].

Programming Languages and Toolchains

Lack of timing constraints in the specification of common programming languages is another problem in the predictability and portability of software with tight timing constraints. For example the specification of C programming language—which is very popular and widely used in the embedded software development—does not mandate timing constraints for the runtime of the generated binaries [7]. The great variety between the processors today makes it much harder for vendors to develop their compilers by considering portability, performance, stability and predictability. Generally, performance is inversely proportional to portability and predictability. Performance (runtime speed, compilation time or generated binary size) is a tradeoff for compilers and mostly it is one of the important reason for optimization. Many optimization types and techniques exists today [8] and compilers also get better results than software developers in code generation by the help of automatic code optimization.

Different types of optimizations like loop optimizations or machine code optimization for a specific platform are allowed as long as the behavior of the executable is compatible with the specification [9]. However, there is nothing about the timing constraints. Toolchain vendors [10]—either commercial or open source—mostly try to optimize performance in terms of speed or size. Although short execution time is mainly the target of optimization by developers speed without control may suffer from the problems of timing and resource management. More drastically, even the systems without any defined timing constraints can suffer from code optimization of the compiler which may cause unpredictable behavior at runtime or even software crash.

Storage Problems

Flash memory is one of the most common and emerging storage technology on embedded systems. When compared with hard disk drives speed, compact structure and mutability to physical damages due to the lack of mechanical parts make flash memories popular in embedded world instead of hard disk drives [11].

In case of a system design, storage with flash memory technology has two main considerations. First, different vendors may have different software interfaces for their products and second, the performance of storage subsystem due to the changing conditions or even between different vendors. First problem is easy to solve and even a well-known pattern like adapter pattern [12] can be used to achieve abstraction of the system during the design phase. However, second case is harder to overcome because of the difficulty of performance and lifetime related problems.

Flash memory write endurance [13] which is directly related with the lifetime is a serious issue for robustness and high availability of a system. Although wear leveling [14, 15] seems to be a solution to this problem it does not promise much for the robustness of the system. Moreover, the algorithms [15] used in wear leveling may decrease the overall performance of the flash memory which is another problem for the system designers. It is possible for different vendors to implement different wear leveling algorithms [16] which results with different effects on performance. Although, vendors publish the specifications of their product most of the time these are not meet the requirements of such complex systems and promise much about predictability. Moreover, physical condition in the field like temperature may directly affect the storage subsystem. Hopefully, most of the time it is possible to formulate and model the effects of environmental conditions to storage subsystem and by the help of calibration methods it is quite feasible to adapt a system to a new environment. In a fault tolerant system error detection and correction still depends on test which may be done on demand or periodically to detect errors on the storage subsystem. The results may be sent to a central unit in order to warn the person in charge or used autonomously to recover from this situation in order to achieve robustness. Reserving different memory areas for the same purpose or even different flash memory banks to backup ongoing system and achieve high availability are seems to be possible and also simple solutions. Such a system seems to be possible, but from the point of a software developer it is much more difficult to cope with hardware failures than software failures [17].

Cache Problems

Cache memory is one of the fundamental parts of modern computer architecture today. A well designed application can benefit from cache in order to speed up its execution time by decreasing the time spend for memory access to fetch required data or instruction. However, in real time systems cache memories is a problem for predictability and sometimes it is a challenge for system designers to overcome.

Most challenging problem is Cache Interference Cost [18, 19] in a multitasking environment which is conventional method of software development today. Context switching between tasks makes the current state of the cache unusable for the next running task. While calculating the Worst Case Execution Time (WCET), in addition to context switching times, Cache Interference Cost should be considered to get realistic results from the analysis.

Dynamic scheduling in multitasking systems makes harder to do scheduling analysis when cache memories are considered. However, preemptive scheduling may be the best option in order to meet the requirements. Cyber-physical systems or even today's simple embedded devices need more communication than in the past. Not only between each other, but also frequent communication with users is an unpredictable event in the lifetime of such a system. For example a device which uses network sockets over a local area network has to response user requests in an acceptable time period. Here assume that task A is responsible for data acquisition from sensors periodically and task B is responsible for executing user requests. Task A is periodic, but task B sporadic and it is impossible to know when it will be active in the system. Preemptive priority based scheduling is an option here to schedule the execution of these tasks and meet the requirements. Let's give task B higher priority than task A to make it the running task in the system when a user request has arrived. In this scenario when user send a request task A should be interrupted to make task B the running task. Because this cause a context switch, the state of the cache memory which had been used by task A will be irrelevant for task B and in that case a cache miss triggers a Cache Interference Cost. Because it is quite hard may be impossible sometimes to know the context switching in advance, it is quite hard to analyze the effect of cache interference cost on predictability. Let C is the time for the worst case execution time without context switching (S) and cache interference cost time (I). C' is the total of C plus context switching times (2S) and cache interference time (I), which can be formulated as $C' = C + 2S + I$ [18]. Context switching and cache interference cost shift the estimated time C to C' and because S and I are not constant for every case and platform, these items decrease the predictability of the systems. In addition to predictability, performance of the system also suffers from this case.

Simplest solution to this problem is disabling caching and losing the benefits of cache memories totally. This tradeoff makes sense when predictability and timing are the main non-functional requirements of the system. However, without compromising the benefits of cache memories we can use them use them after the eliminated the cache interference cost factor. Cache partitioning is an alternative solution to this problem which is widely used today [20]. Cache partitioning also enables the isolation of tasks in an environment with shared resources like CPU, cache and buses. There are many portioning techniques exists today to benefit from cache memories, but in basic terms cache partitioning is based on the idea of assigning reserved partitions of cache memory to tasks in order to achieve the isolation mentioned previously. Cache locking is another solution which depends on the idea of locking the cache memory in order to improve the timing predictability and WCET analysis [21].

Network Problems

Communication is one of the building blocks of cyber-physical systems. Communication may occur between human - machine or machine—machine. Wide and local area networks are widely used for conventional networking. This data oriented usage of networking has different requirements than the real-time systems which depend on reliability and predictability. Basically, loss-free or even just high speed data transfers are the basic problems to solve for conventional networks. Real-time systems focus on reliability and predictability than speed or high bandwidth. Of course high network utilization, high data bandwidth and less complexity are also demanded properties for real-time communication, but conventional networks suffer from more serious problems than these.

In terms of predictability, variation of data transfer delays (jitter) over a network is an extremely unwanted and dangerous factor for real-time communication [22]. Delays, if they are accurate which means that minimum and maximum delay times in a network are quite close to each other, are easy to handle by using different methods like buffering [23]. Moreover, circuit-switched networking can achieve very good results for delays. However, they are note enough to struggle with jitter problem within today's complex networks.

Integration ability to other systems is another aspect of communication between cyber-physical systems [23]. Especially communication between systems is a key achievement to create systems adopting, learning and even teaching. An air conditioning system which queries the daily weather information from your television and inform your refrigerator the current degree of your home set by you. This is a huge collaboration without any human interaction. Therefore, it is important to make this process autonomous in order to increase usability and interoperability of these systems. Commercial disputes between companies and different requirements of each domain make creating homogeneous networks harder in terms of standardization, but open standards and efforts for interoperability is offering great benefits.

Because a network between cyber-physical systems can consists of many nodes, the efficiency and required processing power on the nodes should satisfy the requirements of cyber-physical systems. Delays, jitters on the network, network traffic due to broadcasting, multicasting messages and power requirement of network devices make deploying networks harder.

Concurrency Problems

Multitasking and increasing demand for parallelism popularize the use of threads almost in all projects today. Although multithreaded systems are quite difficult to analyze and implement, multithreaded applications is the most common way of achieving parallelism. Deterministic sequential program execution is corrupted by the threads and it becomes a nondeterministic problem to be solved. Order of sequence, shared resources and deadlocks are main concerns when dealing with the non-determinism of multithreading. Locks, semaphores, monitors and similar solutions exist in order to manage this complexity [24]. However, multithreaded programming is still a hard and error prone task. Although, locking the critical regions in the same order between the threads is a well-known solution to deadlock problem, most of the time it is hard to preserve this order between threads in different piece of code.

Conclusion

This chapter addressed some of the design problems and challenges to migrate our knowledge on real-time embedded systems to cyber-physical systems. Although, balance between our capabilities and increasing complexity of cyber-physical systems, it is a great challenge to shift a new paradigm. Current software development methodologies have problems coping with large scale integration of cyber-physical systems, increasing demand for more processing power and communication between these systems. Cyber-physical systems is about to open a new era in the history of computing and the tools we have are still belongs to an older age. Although current methodologies are also promising, it is still uncertain that how long we can continue with the solutions we choose to prolong the lifetime of current methodologies and technologies. Because collaboration is in the nature of cyber-physical systems, collaboration between different fields of science is necessary for further achievements.

References

1. C.L. Liu, James W. Layland, "Scheduling Algorithms for Multiprogramming in a Hard Real-Time Environment", Journal of ACM, vol 20, pp 46–61, 1973.
2. Timo Mantere, Jarmo T. Alander, "Evolutionary software engineering, a review", Applied Soft Computing, vol. 5 issue 3, pp. 315–331, 2005.
3. Reinhard Wilhelm, Jakob Engblom, Andreas Ermedahl, Niklas Holsti, Stephan Thesing,David Whalley, Guillem Bernat, Christian Ferdinand, Reinhold Heckmann, Tulika Mitra,Frank Mueller, Isabelle Puaut, Peter Puschner, Jan Staschulat, Per Stenström, "The Worst Case Execution Time Problem - Overview of Methods and Survey of Tools", ACM Transactions on Embedded Computing Systems (TECS), vol 7, Article No 36, 2008.
4. Heiko Falk, Paul Lokuciejewski, "A compiler framework for the reduction of worst-case execution times", Springer Real-Time Systems, vol. 46, 2010.
5. Sharad Malik, Margaret Martonosi, Yau-Tsun Steven Li, "Static timing analysis of embedded software", DAC '97 Proceedings of the 34th annual Design Automation Conference, pp. 147–152, New York, NY, USA, 1997.
6. Christian Prehofer, Jilles van Gurp, Jan Bosch, "Compositionality in Software Platforms", Emerging Methods, Technologies and Process Management in Software Engineering, Wiley, 2008.
7. E.A. Lee, "Cyber Physical Systems: Design Challenges", Object Oriented Real-Time Distributed Computing (ISORC), 11th IEEE International Symposium Conference Publications, pp. 363–369, 2008.
8. Keith Cooper,Linda Torczon, "Engineering a Compiler", 2nd edition, Morgan Kaufmann, 2011.
9. ANSI, "ISO/IEC 9899:1999 Programming Languages - C," 1999.
10. Intel Corporation, "Quick-Reference Guide to Optimization with Intel Compilers," version 12, 2010.
11. B. Dipert, L. Hebert, "Flash memory goes mainstream," IEEE Spectrum, vol. 30, pp. 48–52, 1993.
12. E. Gamma, R. Helm, R. Johnson, J. Vlissides, "Design Patterns: Elements of Reusable Object-Oriented Software," Addison-Wesley Professional, 1994.
13. S. Boboila, P. Desnoyers, "Write endurance in flash drives: measurements and analysis", FAST'10 Proceedings of the 8th USENIX conference on File and storage technologies, 2010.
14. Ken Perdue, "Wear leveling application note," Spansion Inc., 2008.
15. Li-Pin Chang, "On efficient wear leveling for large-scale flash-memory storage systems, "SAC '07 Proceedings of the 2007 ACM symposium on Applied computing, pp. 1126–1130, 2007.
16. Eran Gal, Sivan Toledo, "Algorithms and data structures for flash memories," Journal ACM Computing Surveys, vol. 32, issue 2, pp. 138–163, 2005.
17. Yuan Chen, "Flash memory reliability NEPP 2008 task final report, " NASA Jet Propulsion Laboratory, 2008.
18. S. Basumallick, K. Nilsen, "Cache issues in real-time systems," AGM SIGPLAN Workshop on Language, Compiler and Tool Support fot Real-Time Systems, 1994.
19. Jochen Liedtke, Hermann Härtig, Michael Hohmuth, "OS-Controlled cache predictability for real-time systems," Real-Time Technology and Applications Symposium Proceedings, pp. 213–224, 1997.
20. M. Caccamo, Sha Lui, J. Martinez, "Impact of cache partitioning on multi-tasking real-time embedded systems," Embedded and Real-Time Computing Systems and Applications Conference Publications, pp. 101–110, 2008.
21. Xavier Vera, Björn Lisper, Jingling Xue, "Data cache for higher program predictability," Proceedings of the 2003 ACM SIGMETRICS international conference on Measurement and modeling of computer systems, pp. 272–282, 2003.

22. Donald L. Stone, Kevin Jeffay, "An empirical study of delay jitter management policies," Multimedia Systems, Springer Berlin, pp. 267–279, 1995.
23. Dinesh C. Verma, Hui Zhang, Domenico Ferrari, "Delay jitter control for real-time communication in a packet switched network," Communications for Distributed Applications and Systems, Proceedings of TRICOMM, pp. 35–43, 1991.
24. E. A. Lee, "Problem with threads," Computer, vol. 39, issue 5, pp. 33–42, 2006.

Printed in the United States
By Bookmasters